TURQU

Mines, Mine
Wearable Art

Mark P. Block

Schiffer Publishing Ltd
4880 Lower Valley Road, Atglen, PA 19310 USA

Dedication

For the rock hound and collector; miner and mineralogist; artist and craftsman who enjoy the beautiful and precious gemstone turquoise.

Other Schiffer Books by Mark P. Block
Contemporary Marbles & Related Art Glass.
The Encyclopedia of Modern Marbles, Spheres, and Orbs.
Las Vegas Lights. Mark P. Block and Robert Block.

Other Schiffer Books on Related Subjects
Collecting Fluorescent Minerals. Stuart Schneider.
Evolving Southwest Indian Jewelry. Nancy N. Schiffer.
Indian Jewelry of the American Southwest. William A. Turnbaugh & Sarah Peabody Turnbaugh.
Indian Silver Jewelry of the Southwest, 1868-1930. Larry Frank with Millard J. Holbrook II.
Jewelry by Southwest American Indians: Evolving Designs. Nancy N. Schiffer.
Mining in the Old West. Sandor Demlinger.
Southwest Silver Jewelry: The First Century. Paula A. Baxter.
Turquoise Jewelry. Nancy N. Schiffer
The World of Fluorescent Minerals. Stuart Schneider.

Library of Congress Control Number: 2007921859

Designed by John P. Cheek
Cover design by Bruce Waters
Type set in Bernhard Modern BT/Souvenir Lt BT

ISBN: 978-0-7643-2642-4
Printed in China

Published by Schiffer Publishing Ltd.
4880 Lower Valley Road
Atglen, PA 19310
Phone: (610) 593-1777; Fax: (610) 593-2002
E-mail: Info@schifferbooks.com

In Europe, Schiffer books are distributed by
Bushwood Books
6 Marksbury Ave.
Kew Gardens
Surrey TW9 4JF England
Phone: 44 (0) 20 8392-8585; Fax: 44 (0) 20 8392-9876
E-mail: info@bushwoodbooks.co.uk
Website: www.bushwoodbooks.co.uk
Free postage in the U.K., Europe; air mail at cost.

Contents

Acknowledgments

Rainbow

Shimmering color arched against grey sky,
Painted by dancing light on air-borne mist.
Wide flung by a sacred hand...
The Hand that formed of dust nothingness
The solid Earth below.
Beauty and promise together blended,
Beauty ethereal, promise divine.
Given to grace the clouds and the rain,
Given to bless the world-weary heart...
Shimmers... fades... brightens...
To vanish in brilliance...
Shines through the dark in my soul.

This poem, *Rainbow* by Red Unicorn (Barbara Mann), provides an appropriate grounding for this book focusing on turquoise, its mines, and wearable art. Throughout the blue and green, yellow and tan colors turquoise can be found, in which the matrix colors of brown, black or even red, offer a rich rainbow of their own. As the words of the poem come together to tell a story I have a sense of being transported to another region, one that has always fascinated and enthralled me. It has done so through the people I have met and known over the course of 30+ years as much as by the landscape, art, and culture the American Southwest has to offer. [Figure I]

Figure I. Various single and multiple stone cluster turquoise and silver cuff bracelets. These bracelets include Chinese, Kingman, and Sleeping Beauty mine turquoise. The bracelets are offset by a host rock with natural turquoise veins clearly visible. Pueblo and Navajo. 1-3/4" - 2-1/4" x 3". *$500-800. Courtesy of John Renner.*

My first piece of turquoise came to me by postal mail many years ago. It was in amongst a coffee can of agate, chalcedony, jasper, and other minerals. Sent to me by a desert dweller, Marguerite Knickerbocker, first a long-distance pen pal and shortly thereafter a dear, sweet friend from the sleepy ghost town of Nelson, Nevada. No more than thirty people back in the mid-1970s inhabited this town. Nelson is nestled at the base of the Eldorado Canyon in Southern Nevada, about forty miles south of Las Vegas, and served as a hub of activity in the late 1800s surrounded by some of the greatest silver mines of the United States, including the famed Techatticup and Wall Street Mines. Its history is legendary as the most violent town of the old West, yet the inhabitants of the late twentieth century were some of the finest people you could ever want to meet, even if the sign by the road "in town" stated, "Don't Even Think About Stopping Here." [Figure II]

Figure II. Nelson, Nevada, ca. mid-1970s. *Courtesy of the author.*

I owe a debt of gratitude and everlasting love to my late friend and "grandmother" Marguerite Knickerbocker. Her ability to withstand the harshness of winter and the searing heat of summer was overshadowed by her loving and protective nature. She often told her friends, both local and celebrities, that no one could find "such beautiful rocks" laying on the ground or on the hills and mountains of the canyon quite like I could. My eyes were always looking down, and I delighted in showing her my treasures after each of my hikes. Without doubt she would have loved this book.

Aldina Mang, Bill Hirst, Bob Hill, Carmen Methven, and a host of Western "old-timers" enhanced my love and appreciation for the American Southwest over the years. Sadly, all have now passed on and are at peace, yet every time I travel to the region I feel their presence and recall with affection the days and nights of a much quieter time. A time when Las Vegas and the surrounding area was less crowded, the neon truly vintage, the desert more open, and you could count the hotels and casinos on two hands. Each of these dear friends imparted in me a sense of belonging, and often longing when we were apart.

This book would not have been possible without the exceptional willingness and participation of John Renner, owner of JR's Southwestern Trading Post in New Hope, Pennsylvania. John has a unique perspective on both turquoise as a gemstone and its value as wearable art. Both from his shop in New Hope and through his travels across the United States, John was available, accessible, and excited to work to ensure this volume contained a vast cross section of cut, polished, and set stones, thus enabling the reader to distinguish the various mined turquoise examples necessary to link the understanding of the gemstone with the notion that a single gem could have so many variations in color and appearance. He is a wealth of knowledge on the subject and an expert at identifying turquoise mines by simply viewing the stone. John, you have my enduring thanks, appreciation, and friendship. A special thank you to Barbara and JoAnne as well for their aid in reaching the printing stage. Except where otherwise noted in each illustration caption, all pieces are *courtesy of JR's Southwestern Trading Post*.

With each book I have written and been fortunate to have published, Peter and Nancy Schiffer have both extended their hands in friendship and their company's willingness and faith in the particular project. They epitomize the definition of friends, and along with the entire staff at Schiffer Publishing, their encouragement and professionalism have given me hours of joy and the energy and fortitude to see the process to its bound conclusion.

A special thank you is due Doug Congdon-Martin of Schiffer Publishing. I would have hardly thought it possible to squeeze camera equipment, himself, stones, jewelry, and even me all into one 10' x 10' room on site in New Hope. Yet, Doug never once complained and produced some of the most beautiful images of turquoise jewelry you could imagine. Now that is true professionalism, as is that which my editor Jeff Snyder once again brings to this book project.

No one should write any book without the support and love of their family. I have been more than fortunate in this regard to have both. I have been blessed with a loving wife and daughter who enjoy traversing the country to aid in photo shoots, research gathering, and appreciating the beauty of minerals and art glass, antiques and collectibles. Each book project has been a challenge to meet and a reward to read. It is all the more enjoyable to have them with me on life's journey. My love to you both.

Foreword

Working with turquoise and being in the Indian turquoise jewelry business has been an amazing adventure, an adventure that has led me to aid in this important project—documenting the various mines and varieties of turquoise they offer the public.

I have been on a quest to search and find beautifully exotic older turquoise jewelry, pieces that exemplify the incredible diversity of various Native American traditions and artists since my earliest days in the business. Equally as interesting for me has been showcasing, educating, and selling these beautiful pieces to new owners.

Over the years I have found great joy in cutting, polishing, and setting turquoise in sterling, and occasionally in gold, in order to create objects of spirit and beauty. I created and staffed a lapidary and silversmith workshop in Albuquerque, New Mexico, and spend most of my waking hours involved with turquoise in one way or another, either set in jewelry or in the rough. I either buy it, sell it, or make it into jewelry. There has been no better education in identifying and appreciating the immense variety this beautiful gem has to offer. [Figures III-IIIa]

Figures III-IIIa. New Mexico landscape. *Courtesy of John Renner.* Various single and multiple stone silver cuff bracelets. These bracelets include turquoise, coral, sugilite, and lapis. Signed: John Renner. 1-1/2" - 3-1/2" x 3". *Courtesy of the artist. $500-1100.*

As a jeweler and collector with an appreciation for high quality turquoise, I am always on the lookout for the best available. The availability drops and the prices rise each year. Because I own a store, JR's Southwestern Trading Post in New Hope, Pennsylvania, and am often traveling to shows across the country, I don't have to limit my buying to just old or just new turquoise jewelry. I buy what I find personally interesting. [Figure IV] When people come up to my show booth at one of the indoor antique shows where I exhibit they often ask where I got the amazing quantity and assortment of turquoise jewelry I have displayed. I tell them, estates sales, pawnshops out West, flea markets, the occasional wholesaler, private sales, some I design and have made, and a few pieces I build myself. Each piece is purposefully unique and has its own beauty. [Figure V]

Figure IV. Display. JR's Southwestern Trading Post, New Hope, Pennsylvania. *Courtesy of John Renner. Photo by the author.*

Figure V. An exciting display of various types, styles, and forms of turquoise jewelry from JR's Southwestern Trading Post on exhibit at the Miami Beach Antique Show, Miami, Florida. *Courtesy of John Renner.*

Because many of the turquoise mines in America have closed or have limited production, for most people the best and most affordable way to get good quality American turquoise is in buying pre-1990s native American jewelry. However, the supply has difficulty keeping pace with the increased demand, and so less and less is available from year to year. When buying turquoise jewelry remember there are occasionally great bargains to be had, but as in life usually you get what you pay for.

The recent dramatic increase in the price of silver was a bonanza for metal scrappers but a tragedy for lovers of turquoise and Indian silver jewelry because so many older jewelry pieces were scrapped. This also happened in the 1970s when the Hunt brothers tried to corner the silver market and drove the price of silver to $50 per ounce.

The contemporary turquoise jewelry market is dominated by imported turquoise, primarily from China. Good quality Chinese turquoise is very beautiful and is often difficult to tell from some of the best American turquoise. This turquoise is increasingly being used for contemporary Native American jewelry. If you want to collect American turquoise or American turquoise jewelry, the best piece of advice I can give is know who you are dealing with and spend some time learning about turquoise itself. It is sometimes difficult even for experts to identify a particular piece of turquoise. While there are identifying characteristics like color and matrix for individual mines, a mine can and often does produce some rocks that look closer to the characteristics of another mine. I have 50 lbs of natural Cerrillos turquoise with thick turquoise veins in large pieces. In a few rocks you can see on one side turquoise that is a beautiful sky blue and on the other turquoise that is dark emerald green. I would say seven out of ten turquoise stones are easily identified. The more turquoise you handle, the better you get at identifying the gem's origin. A reputable dealer or knowledgeable collector not only gives you an informed opinion but also helps further your education. They'll stand behind their opinion with a money back guarantee. That dealer recognizes the grade and mine of origin of a piece of turquoise and knows its fair market value.

Turquoise jewelry is not just for celebrities, although plenty wear it and are also collectors. For thousands of years turquoise jewelry has been a fashion statement and was believed to have healing and protective powers. Personally, I always carry a piece of turquoise on me. You could call it habit or superstition but turquoise has called my name my whole life. A Navajo medicine man recently said to me that I didn't choose turquoise—turquoise chose me. He said my young life was a journey towards turquoise and my adult life has been a journey with turquoise. I am grateful to the core of my being for the great blessings that have come my way, the knowledge, the experiences, and the friendships, all over a blue gem of beauty—turquoise. I wouldn't trade the adventure for anything in the world.

John Renner
New Hope, Pennsylvania

Introduction

Turquoise: Prized for Millennia

Look up at a clear blue sky and you'll see quite vividly what has attracted people to the gemstone turquoise for millennia. Whether the sky is light blue in spring, deep blue in summer or even a grayish blue in winter, each season is unmistakable in nature, just like the mineral prized by so many for its own natural beauty.

There is no other mineral that is more beloved for its robin's-egg blue color as high quality turquoise has come to be regarded—the "Tiffany of Gems." In fact, there is no other mineral that exhibits this color. And while this distinctive blue may be the color most are familiar with, there is in fact a myriad of colors within the blue/green range that turquoise exhibits. [Figure VI]

Much of what has been written over the years has been about turquoise as a gemstone, its history from ancient times to the present, and the influence of Native Americans in particular in making it a part of our popular culture. Images of this beautiful mineral have been shown in its natural state, cut and polished lapidary form, and silver and gold set pieces. This volume is intended to bring you closer to numerous types and styles of turquoise mined, cut, and set into jewelry as wearable art—more than you may have ever anticipated or seen before.

Imagine traversing the globe from ancient Persia through Mexico, from North America to China…the path of turquoise will take you on an incredible journey of variety and beauty. For each region is distinctive in what

Figure VI. Morenci mine, left, and Manassa mine, right, turquoise pendants set in silver. Navajo. 1-1/2" x 1-3/4", 1-1/2" x 1-1/2". *Courtesy of the author. $175-350.*

Figure VII. Natural Kingman mine turquoise. Forty-two grams. *Courtesy of the author.*

Figure VIII. Royston mine turquoise, rough, and cut and polished. Thirty-three grams. *Courtesy of the author.*

its mines produce. In fact, each mine can be distinctive in and of itself depending on the vein worked. [Figure VII] You will be dazzled by the variation in color that turquoise exhibits, as well as the manner in which it has been set in silver and gold as it has been mined from both the Southwestern United States, and other localities around the world. [Figure VIII]

Would it surprise you to know that as a mineral turquoise would have little to entice excitement without its distinctive range of colors? It is fairly often described as a dull mineral. It is an opaque gemstone and not translucent like rubies, sapphires, or emeralds, yet it has held man's attention and respect as a prized gemstone for thousands of years. In fact, the very term "turquoise" has taken on a meaning in culture as a recognized shade of blue all its own. Adding to the history and importance of this gemstone, it is believed that turquoise was set into the breastplate of that worn by the high priest Aaron in biblical days. Its hold over people and fascination was enhanced when the treasures of the Egyptian pharaoh Tutankhamen revealed a significant amount of some of the most beautiful turquoise ever seen. Clearly, turquoise was prized for both its beauty and value to the most important leaders of the Egyptian people. Turquoise continues to hold the uniquely historical and distinctive honor of being placed among the finest gemstones known to man. Turquoise can be enjoyed by anyone interested in collecting the rough natural stone, stones that have been cut and polished or those set in wearable art.

Highlighting the text of this book are hundreds of illustrative images that compare various mined turquoise pieces with each other. You are about to undertake a wonderful journey into one of nature's most impressive gemstones. At the same time this book should serve as an important reference and resource in identifying the pieces in your own collection, in addition to those you would like to add. After all, the characteristics of turquoise are both unique and personal to the source of its discovery.

Because relatively little has been published on the turquoise mines themselves, let's turn the tables and see what awaits us as we focus on mines of exceptional reputation like Cerrillos, Sleeping Beauty, and Kingman, as well as some that are more obscure or little known like Ajax, Blue Jay, and Easter Blue. Each will give up a secret gemological treasure that, when cut and set both by native Americans and contemporary jewelers, will at times overwhelm the senses. Yet, more importantly, exploring these mines will document the importance of many of these now lost or depleted mines to the annals of mineralogical history.

Mark Block
Trumbull, Connecticut

Author's Note

The author has compiled the information contained in this book from what are recognized to be authoritative sources and believes that it is both accurate and factual as of the date printed. All written information and illustrations are offered only as a guide to the reader concerning the subjects mentioned. The author makes no representations regarding the interpretation that may be obtained by the reader. It shall be the responsibility of the reader to determine the suitability of any information mentioned.

All copyright and trademark works illustrated and discussed in this book are the properties of those that own them. No attempt to infringe on any copyrights and trademarks has been made in this text. This book is presented for the education and entertainment of the readership.

Chapter 1
The Mineral

Turquoise is classified as a hydrous phosphate, falling in the same category as apatite, mimetite, pyromorphite, and variscite, to name a few. Formed as water trickled through a host rock nearly thirty million years ago, deposits containing copper, aluminum, and zinc were left behind to create the beautiful gemstone turquoise. [Figure IX]

Figure IX. Blue Gem mine turquoise, cut and polished. Eight grams. *Courtesy of the author.*

There is one important element, a mineral in its own right, that dictates the degree of blue turquoise contains, and that is the mineral copper. The copper content in turquoise is responsible for its color. This color can range from a pale powdery blue to dark blue, blue-green, a matte green, and dark green coloring—all due to the amount of copper in the minerals makeup. [Figure X] However, if the mixture of minerals is such that a greater amount of aluminum was added, then the color tends to be more green or white. [Figure XI] If zinc is added to the mix in greater amounts, the color becomes a yellow-green and the stone itself becomes even harder. [Figure XII] All of this is a natural occurrence in the formation of turquoise.

Figure X. Natural copper specimen, United States. 2-1/2" x 1-3/4".

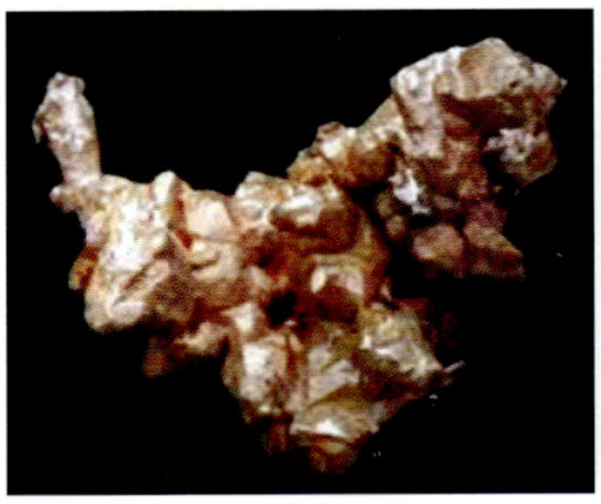

Figure XI. Natural aluminum specimen, Russia. 1-3/4" x 1-1/2".

Figure XII. Natural zinc and sphalerite specimen, United States. 3-1/2" x 2-3/4".

When other minerals appear in the host rock that turquoise was formed within, they are referred to as matrix. Matrix in turquoise rock can be seen as spider web looking or in larger blotches. Spider web is matrix that is most often fairly evenly spaced or having an almost design-like appearance upon the surface of the stone. It will often enhance the value and collectibility of the cut and polished turquoise. A matrix of black is often caused by the introduction of pyrite to the stone and yellow to brown matrix is caused by the introduction of iron oxide. [Figure XIII]

Figure XIII. Easter Blue mine turquoise. Pendant. Silver setting. Navajo. 2-1/4" x 1-3/4". *$200-325. Courtesy of the author.*

From the romance language of French, the word "turquoise" is commonly translated to mean "Turkish." This is of interest not because of the minerals color, but rather because turkeys and turquoise were both introduced to Europe through Turkey. The Middle East of old was a major producer of turquoise and Persian turquoise is consistently sought out for its absolutely stunning pure blue beauty. Just the name conjures up images of beauty and romanticism. [Figure XIV]

Figure XIV. Morenci mine turquoise and black onyx set in sterling silver. Three-level custom designed ring. 1-1/2" x 1" x 1-1/2". Zuni. *Courtesy of the author. $350-600.*

Dating the first discovery of turquoise is extremely difficult. Unlike Sutter's Mill, where the discovery of gold in the American West is well documented, it is not known when turquoise was first identified as a gem of appeal. What is generally recognized is that turquoise appears to date by 5,000 years "Before Common Era." It is known that four bracelets belonging to Queen Zar and found on her mummified remains date to the second ruler of Egypt's first dynasty and occurred over 5500 B.C.E. (Before the Common Era). As more and more discoveries were made in Egypt, turquoise beads were found in tombs and as personal adornments. Later dates indicate turquoise was inlaid with gold, the precious metal used extensively for kings and queens of the region to craft jewelry of the highest form by the finest craftsmen working for the rulers of their time.

Further, it has generally been acknowledged that the oldest known source of turquoise were the Maghara Wadi mines located in today's Sinai Peninsula. Thousands of laborers worked these mines and the stones were meant for the pharaohs, proving the mining operations worthy of working for over 2000 years. Lost to time, the mines were rediscovered in the middle part of the nineteenth century and occasional work took place at them up until the early twentieth century. These mines have since been depleted and there is no active work occurring at them today, though if you know your turquoise well enough it is possible to find a piece of this rare material even in the early twenty-first century.

With the passing of time it is known that turquoise was found in many regions of the world. This is true because archeological history points to the Aztec kings having worn turquoise. It is unlikely that trade accounted for this taking place across the globe from the earlier Pharaohs of Egypt. Not only is the time not correct, but also the trade route passages would indicate this to be highly unlikely. The pre-Columbian Indians had beads and pendants made from turquoise. The color was so distinguishing that the name turquoise has been used to describe any color that resembles it at all.

Turquoise first began to appear in North America around 200 B.C.E. and was used in many of the same ornamental ways as those used further to the South. These Southwestern natives and Indian tribes of Mexico produced beautiful beads and necklaces from locally mined turquoise, a precursor of the explosion in turquoise jewelry intended solely for ornamental use that would take place later on and continue nearly unabated to the present day.

There are ruins and other evidence that indicates quite substantially that turquoise was mined by both the Hohokam and Anazasi peoples. The material would have been found at the famous mines of Cerrillos and the Burro Mountains of today's New Mexico in the American Southwest. Other material has been identified as coming from Kingman (Arizona) and Morenci (Arizona), two of the most famous turquoise mined areas in North America. The America's proved to be a fruitful trading ground for turquoise as it is known that Cerrillos turquoise has been found hundreds and hundreds of miles from it's mines into Mexico where it was used by the Aztecs. [Figure XV]

What is often discussed and many conjure up when thinking of turquoise is relatively modern. These are the works crafted with turquoise and mounted in silver by Native American's—popularly known as Indian jewelry. It was the Spanish who first brought their knowledge of silversmithing to the Southwest. Combined with the lapidary knowledge of the Native Americans, an industry was born. [Figure XVI] This industry remains quite separate from the religious and personal adornment pieces crafted by the Native Americans, and one that is a vigorous income-producing source for tribes, including the Navajo, Zuni, and Hopi. [Figure XVII, XVIII]

Figure XV. Cerrillos mine turquoise set in a silver cuff bracelet. 2" x 3". *$400-550. Courtesy of John Renner.*

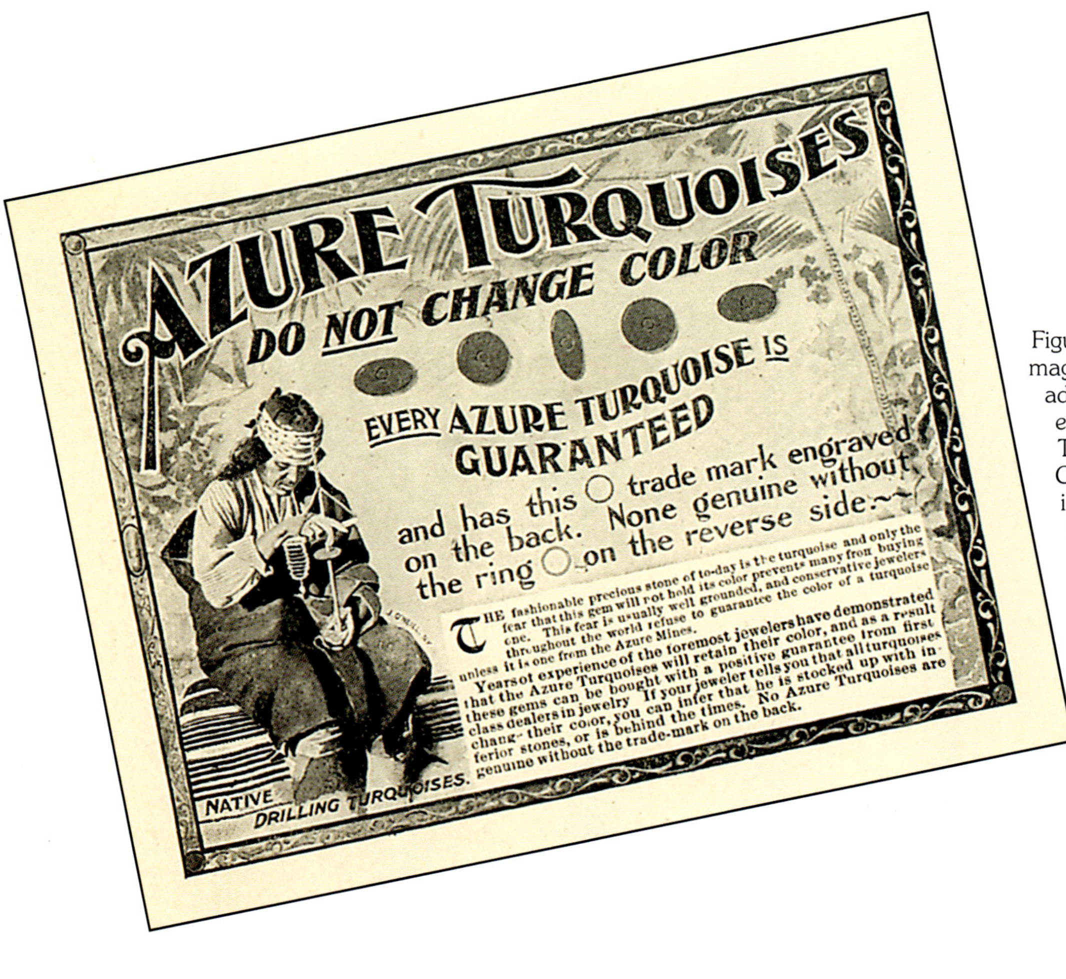

Figure XVI. Azure Turquoises magazine advertisement: This advertisement appeared in the early 1900s and stated, "Azure Turquoises Do Not Change Color. Every Azure Turquoise is Guaranteed and has this O trade mark engraved on the back. None genuine without the ring O on the reverse side." 4-1/2" x 3-1/2". *Courtesy of the author. $15-25.*

Figure XVII. Vintage postcard illustrating turquoise drilling. Front reads: *Indian Drilling Turquoise, Pueblo of Zuni, Arizona.* Back reads: *The Zuni Pueblo Indians are experts in making strand necklaces of shell and turquoise, some are silversmiths of merit and make many objects from American or Mexican coins such as rings, bracelets, belt buckles, etc., which find their way to tourists.* 5-1/4" x 3-1/2". *Courtesy of the author. $5-15.*

Figure XVIII. Vintage postcard illustrating turquoise gem cutting. Front reads: *Largest Gem Cutting House on the Pacific Coast. Southwest Turquoise Co., 113 N. Broadway. Los Angeles, Cal.* 5-1/2" x 3-1/2". *Courtesy of the author. $5-15.*

This industry was begun in earnest in the late 1800s for the tourist trade in a region of the United States that was undergoing a vast transformation. The stages and trains bringing people from the East to settle out West often stopped along the way, and it was in these locations that trading posts, in many ways very different from the solely functional posts of earlier days, serviced the "tourists" with native jewelry. Many of these antique pieces continue to appear in collections or at shows as the natives often pawned them for money for life's necessities. [Figure XIX, XX]

XX. Cerrillos mine turquoise in coin silver setting, ca. 1910-1925. 1-1/2" x 3". *Courtesy of the author.* $150-300.

Figure XIX, Assorted vintage 1930s turquoise and silver bracelets. These include Persian, Cerrillos, and Blue Gem mine turquoise. 1-1/2" - 2" x 3" - 3-1/2". *Photo courtesy of John Renner. $175-400.*

This jewelry is often referred to as "coin silver" turquoise, as it was crafted using the silver of melted down coins. It is believed by many that this first occurred about 1880 when a white trader used the artful sense of persuasion on a Navajo craftsman to make turquoise and silver jewelry. This is likely, because prior to this time nearly all Native American jewelry was made up of solid turquoise beads, inlaid mosaics, carvings, or other three-dimensional designs, few of these if any actually using silver. Quite simply, the Native American's had no access or ability to obtain coin silver. The vast silver mines of the West were being depleted about the same time, and silver bullion was of a worth vastly more than what would have been considered for use in Indian trinkets.

We don't often think of the importance turquoise played in the lives of the Native Americans. However, it is important to remember that many things had value to the people of the Southwest, much of it for spiritual purposes and not for ornamental reasons. The jewelry made by Native Americans for spiritual use was quite often more elaborate and beautiful than that which was crafted strictly for ornamentation. This occurred more prevalently before the burgeoning market of today for extremely high crafted silver and gold settings and finely detailed intarsia pieces. Much of the work being done by Native Americans of the Southwest these days can truly be considered fine art. This is especially true today when differentiating between fine silver and gold jewelry of turquoise and other stones. It does not take a well-trained eye to differentiate between what is truly crafted as art, and what is done in a much more simplified manner. To see the differences between "old" turquoise and contemporary artistry, one merely needs to look at the intricacy of the piece and the contemporary flair of the work, as well as the artist's marking. [Figure XXI]

The artful use of turquoise is illustrated quite beautifully through the Native American's belief that turquoise, the stone itself, brought together the spirits of the sky and sea on land, blessing both hunters and warriors. Rare as they are to find, and even though turquoise is a rather porous mineral, a carved turquoise arrowhead was believed to ensure an accurate aim by its holder. The Navajo Indians believed that turquoise cast upon the waters of a river along with prayer would cause rain to fall. So, while it is important to realize that though non-Native American's are enamored with the art of turquoise jewelry and have been for over a hundred years, beginning with the quaint trading post tourists, turquoise is a meaningful gemstone, and therefore continues to hold very special and significant importance, to the Native American people.

Figure XXI. Candelaria mine and Kingman mine turquoise, necklace and matching earrings. Silver setting. Zuni, signed: Ric Laselute. 18" length. *Courtesy of the author*. $500-1000.

Chapter 2
Gem of Beauty

Whether the turquoise is the solid blue of Bisbee (Arizona) compared to the intense greens of Cerrillo's (New Mexico) or the pale blue/greens of #8 (Arizona), each variation has its own innate beauty. To compare the finest Lander's (Nevada) turquoise with that mined in China is both unfair and unfortunate as each mine offers up treasures that are unique in their own right, presenting turquoise with different characteristics for cutting and polishing, carving and setting. Each and every mine has produced turquoise of both historical and decorative value. In fact, turquoise was so highly valued that the crown given by Napoleon I to Empress Marie Louise had its seventy-nine emeralds removed and replaced with Persian turquoise cabochon stones. And turquoise is often imitated with other minerals of lesser value to achieve a similar appearance.

Grades

As was mentioned earlier, the demand for turquoise outstrips the supply of available material for cutting, polishing and setting in jewelry. While this is often the case with fine gems and turquoise is no exception, processes have been developed to aid in satisfying the market.

Long ago it was discovered that lightly colored turquoise, the material of which the vast majority is mined, is both softer and more porous than some higher gem quality varieties. This can vary from mine to mine. By soaking the material in a liquid plastic or polymer, the gem grade mineral deepens in color and hardens as well. Thousands of years ago this process was accomplished through the utilization of animal fat. While the permanence of plastic was not achieved then, the desired darkening of the mineral's color became apparent. Often you will hear the terms "stabilized" and "treated" used interchangeably in reference to this practice. While the process may be slightly different, the effect is the same, to harden or stabilize soft, porous turquoise. "Enhancing" turquoise is a newer term applied to turquoise that has undergone a process of hardening and intensifying the gem's color through an electrical process. It is important to note that enhancing turquoise does not alter its chemical structure. Therefore, the gem grade mineral that can be cut and polished is perfectly acceptable for use in jewelry and is accorded similar status to natural turquoise.

Also, because nearly all turquoise is mined from veins, much of the material can be extremely thin, thereby making it impossible to cut and use in jewelry. This deficiency is overcome when a stone is backed by epoxy, it is then referred to as a "doublet," and the cutting and polishing can take place. This backing enables the stone to be protected from cracking, chipping or breaking. It is a thoroughly suitable and often used practice for taking especially rare or extremely beautiful turquoise and ensuring it does not end up on the scrap heap.

Sadly—unlike the fine gem material of ancient Persia, the Mecca of turquoise mining—because of its inferior color, most of the turquoise mined over the last hundred years or so would be relatively worthless if it weren't for the ability to stabilize the material. This process is accomplished in natural turquoise by submerging it in a stabilizing compound, usually using an epoxy-type resin. This process is not to be confused with others, such as heat-treating, synthetic, or composite form. Stabilizing natural turquoise is perfectly acceptable and allows the lapidary and jewelry artist to create works of accomplished beauty from material that would not otherwise be useful in the crafting process. This is not to say that natural turquoise material that has not undergone a stabilizing process is not presently mined and polished, because it is. The mines of the American Southwest, and more recently China, have produced outstanding gem examples of turquoise that have been cut and polished, designed and set in jewelry of the finest quality.

When grading turquoise you should always take into consideration the stone's **hardness** or **density.** Turquoise ranges from two to nearly six on the Mohs scale, its specific gravity also varies but usually is 2.8, like quartz. Therefore, it is often the case that the harder the stone, the better the cut and polished gem. This is especially true with natural turquoise. The stone's **luster** should be both from the turquoise itself and its outward polished appearance. When considering **color**, be aware of the hue and its intensity. This is true whether the color is blue or green, white or tan. When it comes to the **matrix** of the stone, you want to be sure it is pleasing or unusual and that it accentuates the turquoise, rather than distracting from it. Lastly, **rarity** can be of immense importance. The more difficult to find mined turquoise can often drive the market.

The following definitions are important to be familiar with as you search for and collect turquoise, whether as a collector of cut and polished stones, or a wearer of fine turquoise jewelry.

Natural turquoise is exactly that, turquoise which has undergone no alteration to its composition at all. Gem grade stones of natural turquoise are cut and polished just as they have been mined. While they will remain porous, there is a tendency for natural turquoise to change color with time. This is the result of natural wear and exposure to light. [Figure XXII]

Figure XXII. Natural crystalline turquoise specimen. United States. 2-1/2" x 1-3/4". *Courtesy of the author.*

Stabilized or **Treated** turquoise refers to the natural material having been altered through a chemical process to harden the mineral; this is most often done by filling the porous surface with an epoxy or polystyrene chemical. In effect, what occurs is a stabilization or freezing of the color so changes will not occur at a later time and the stone can then be cut and used in jewelry or carvings. This is the most common and often necessary means by which to make turquoise usable as a gemstone. Lapidaries and jewelers use this material and find it a perfectly acceptable alternative to the rare and extremely expensive natural turquoise. Of importance to note is over eighty percent of all turquoise has been stabilized before cutting and polishing. Often times the material will be quite thin, as it is found in veins; therefore, an epoxy backing will be affixed to the piece before cutting so that as much gem grade material can be cut and polished as possible. [Figure XXIII]

Figure XXIII. Blue Gem mine turquoise. Stabilized. Eighty grams. *Courtesy of the author.*

Color Enhanced or **Color Treated** turquoise, is that material which has been chemically altered to enhance the color of soft and very pale natural turquoise. The process is similar to stabilizing turquoise, only in this instance the hardening agent is mixed with a dye. This dye is most often a robin's egg blue color. After the pressure has been applied to infuse the material with the dyeing agent, this turquoise has a plastic appearance. This is the best way to identify turquoise that has been enhanced or treated. It also looks as if it is imitation material because of its blue color and highly polished appearance. Another easy way to distinguish if this is low-grade turquoise is to take a knife blade or open paper clip and scratch the material. An easy scratch is a good indication this turquoise has been enhanced or treated. Higher quality turquoise cannot be scratched as easily with a knife, even though turquoise falls as a six on the Moh's scale of hardness. The value of this turquoise is considerably less than natural or enhanced material. [Figure XXIV]

Figure XXIV. Kingman mine turquoise. Color-treated. Fifteen grams. *Courtesy of the author.*

Reconstituted turquoise is a way to save all the "scraps" of material mined or ground. The turquoise dust, chips, and bits are mixed with plastic resins and compressed into a solid form. This material is acceptable for use in jewelry and carvings, but again, it is important as a wearer of jewelry or collector to know that the material itself is not in its natural state. The color of reconstituted or block turquoise as discussed below is often quite beautiful itself. [Figure XXV]

Figure XXV. Reconstituted turquoise, coral, shell, and black onyx. Cuff bracelet, multiple stone. Silver setting. Navajo. 2-1/4" x 3". *Courtesy of John Renner.* $225-300.

Block turquoise contains no turquoise at all, though through the use of plastics and dyes it is created to resemble natural turquoise. Other minerals can be added to this block form, including pyrite inclusions to give the material the appearance of accompanying matrix when cut and polished. This material is used most often in inlay work and beads. It is common for block turquoise to be mistakenly referred to as reconstituted turquoise. [Figure XXVI]

Figure XXVI. Block turquoise. Earrings, multiple stone. Navajo. 3-1/4" x 1-1/2". *Courtesy of John Renner.*

Imitation turquoise is just that, material intended to look like turquoise but having no physical properties of the gemstone itself. Minerals that accept dye because of porous qualities have been favorite ores to imitate natural turquoise. Plastic has been an important imitation material for years, being used in costume jewelry and beadwork. It can be formed and shaped to imitate natural turquoise, though in most cases, its luster is a much waxier one than natural turquoise. The most often associated imitation turquoise over the years has in fact been another mineral—howlite. This mineral has the look of spider web turquoise when dyed blue. It is also extremely porous and accepts dye quite easily. The novice can be easily confused when looking at dyed howlite versus natural or stabilized turquoise.

Other minerals used to imitate natural turquoise include chrysocolla, amazonite, and variscite. Each of these minerals has found its way by the inscrutable as a cheap way of mimicking turquoise. Additionally, be aware that glass imitations appear on the market that is distinguishable by the high luster they show. [Figure XXVII]

Figure XXVII. Howlite. Dyed. Eight grams. *Courtesy of the author.*

Value

Valuing turquoise is done by taking a number of factors into consideration. Primarily, is the turquoise itself rare? Does it come from a discontinued mine source, or one that produced little material at all? Is the turquoise of exceptional quality, meeting all of the criteria for color, cut, and polish, and setting? These are just some of the questions you need to ask yourself when you are searching for turquoise jewelry, rough, or cut and polished stones.

More stable than currency, fine gems and jewelry have proven to be an important commodity for centuries. Few would argue that in general, gems have proven to be a good investment. In turquoise it is not likely that you will find many bargains. The material is limited in availability, and certainly high quality turquoise is even more difficult to find either through a wholesaler or directly from a retailer with the passing of time.

Part of the allure of turquoise is knowing that it is the oldest gemstone put to use in the United States. The cultural connection to the American Southwest is extremely strong, and many can identify with the romanticism of earlier days involving the American Indian and West. Images that have defined others for generations are difficult if not impossible to forget, and in many ways the Native Americans of today encourage through their art the very images of the Southwest prevalent for so long.

Of course for turquoise to have value it must have visual appeal. The character of the gem must elicit a response from the stonecutter and polisher to the wearer or collector. This response occurs with smaller mine turquoise; for example, the mere bucketfuls of material that came from Landers (Arizona) and the desirable Blue Gem (Nevada) mines. This stands in contrast to the stunning material that was mined for decades from Bisbee (Arizona), Kingman (Arizona), and Sleeping Beauty (Nevada). Each mine produced turquoise of varied quality, yet quite distinctive and sought after by Native American and other jewelers.

The color of a turquoise stone is the essential component in assessing the value of the piece. Whether the turquoise has a matrix of spider web, pyrite or other inclusions can enhance or detract from the piece—though this is often a more personal variable in valuing the stone. The size, cut, polished shape, and quality are important factors in attributing value to any piece, when considering factors beyond its color. Also, the setting the stone has been placed in can enhance the value dramatically. Has it been singly set or done so with other minerals? Is the piece part of a set? For example, a pendant and matching earrings? Is the piece antique or contemporary? And if contemporary, is the artisan a well-known and respected craftsman? And finally, one should always be familiar with and aware of the gemstone's total carat weight. This increases for an individual stone as well as for the total weight of combined pieces. Turquoise, as other gemstones, is measured and sold by carat weight. A unit of measure to remember is five carats per gram. And in today's market, the price of turquoise can range anywhere from fifty cents per carat to well over one hundred dollars per carat. In the captions throughout this book you can find the associated price range for nearly all of the pieces illustrated. This should only serve as one aid for you in assessing the value of your own pieces or collection.

There are five recognized categories to rank turquoise, and the stone's value is in large part a function of its rank. It might surprise you to learn that less than one percent of all turquoise can justifiably be called a "gem." Still, very high-grade stones are nearly gems themselves.

Very High Grade turquoise indicates the matrix pattern, if any, may not be wholly balanced in the stone, or the color is not nearly flawless throughout the piece, though this would be minor in relation to the overall quality of the stone.

High Grade turquoise is hard, though may not be wholly balanced. The stones luster is of the highest quality. Often, it is this turquoise that is used in the finest custom designed jewelry pieces.

Jewelry Grade turquoise has a very good hardness, and is most often not stabilized. Its luster is extremely good and the matrix, if any, is nicely balanced.

Good Grade turquoise will take a good polish. It is pleasing in every respect, though of lesser quality than the higher grades. This turquoise almost always needs to be stabilized in order to cut and polish it, as well as having it hold its color. Lastly,...

Chalk or **Bulk Grade** turquoise is the softest and most porous of all turquoise. It is extremely brittle and must be stabilized and very often heat-treated or color enhanced to make it workable in any form.

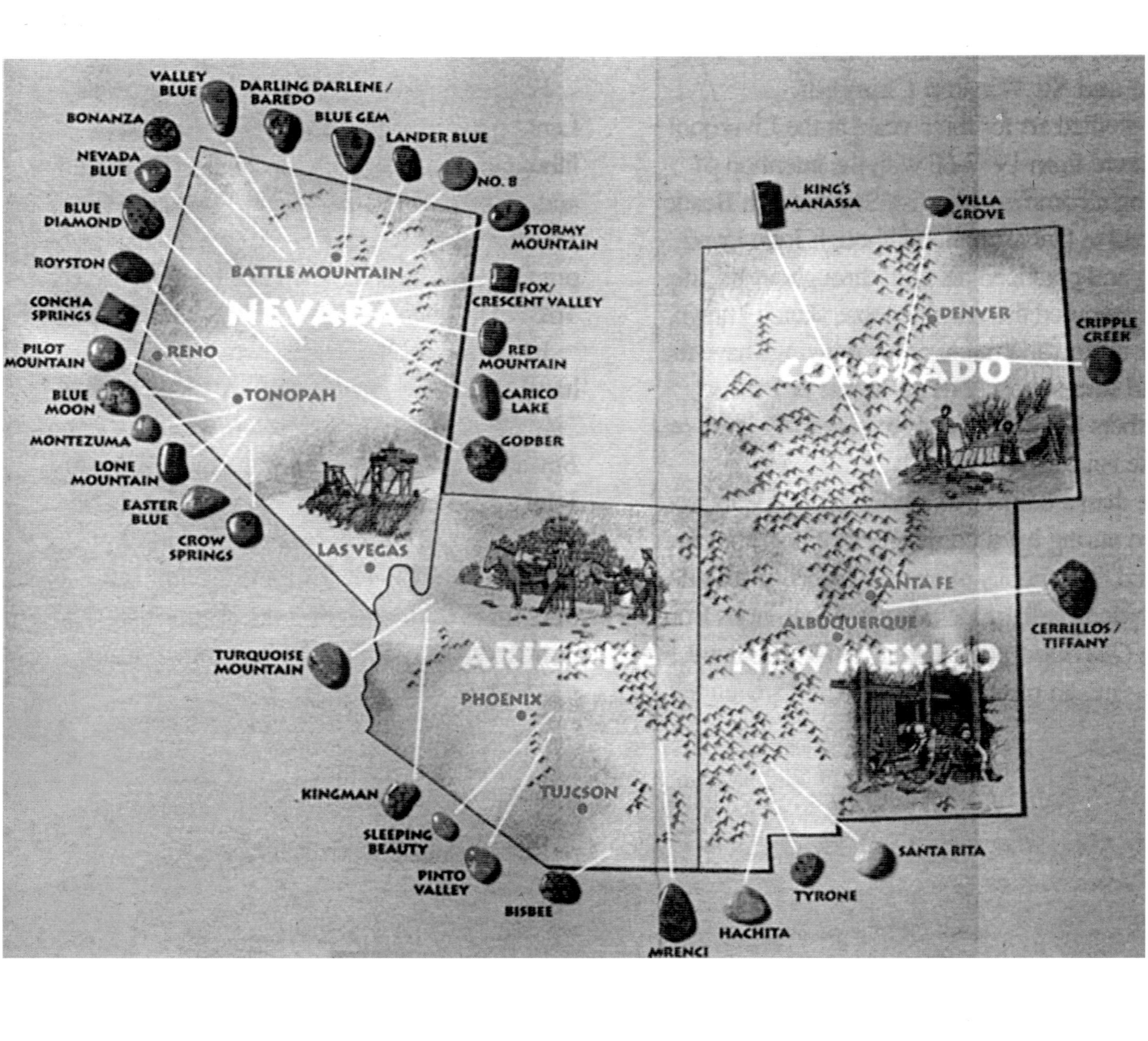

Chapter 3
Countries of Origin

Nation upon nation can lay claim to being a host to turquoise in its native state. While some are extremely well known and accessible, others are rare and often virtually impossible to gain entry to today. A number of countries produced vast quantities of various quality materials, while some nations mined small amounts that account for nothing more than a mere mention. This mention is important however in understanding the worldwide availability of turquoise, and how trading routes came to be important in moving this prized gemstone from one region to another.

An interesting aside is that turquoise appears to have been considered to be a mineral of exceptional beauty and life. Good fortune, and even natural healing powers, have been attributed to turquoise from many disparate cultures around the world throughout the centuries.

While illustrations will follow throughout the text of this book as described earlier, the world map illustrated here [Figure XXVIII], and the discussion of various countries serve as an important set up for understanding the availability and rarity in mining quantities of turquoise for use in crafting.

Figure XXVIII. World map. *Courtesy of the author.*

Egypt has been discussed earlier in a previous chapter as being recognized as the first nation to use turquoise in religious and ornamental manners. The mines in the Sinai were quite active during the rule of the ancient Egyptians, and these people firmly believed that turquoise had powers ranging from healing to mystical. Again, ancient Egypt is acknowledged to be the first to have wearable art and carvings of turquoise going back as far as 5500 B.C.E. Egypt therefore holds a unique "first place" in the history of turquoise as we have known it to be mined and used.

Persia holds a distinction solely different from that of ancient Egypt. The mines at Nishapur revealed the world's finest natural turquoise. The Persians were also the first peoples to use turquoise as a trade item. Bartering their natural resource was an important point of commerce for the Persians and served them well. Ancient Persian turquoise has been found in various areas of Russia, including Siberia, so we know that it traveled across the continent with earlier travelers or settlers, and graves in Turkistan have been opened and exposed some of the finest cut Persian turquoise. Highly prized for its consistent robin's egg blue solid color, Persian turquoise continues to be the standard by which all other turquoise is judged.

Tibet has been a source for some of the most fantastic spider web turquoise for generations. Though the amount of turquoise this country has yielded is small, and some have lumped Tibet and China together, Tibetans valued the gemstone above gold. Not only was turquoise used for personal ornamentation, but it was valued as a currency in many regions of the country as well.

China is today's leading producer of gem quality turquoise. Dating well back to the thirteenth century, most of the early Chinese turquoise came from trade with others. In part, because of the relatively closed society the Chinese have, we are aware of little from individual mines in China. While the turquoise can be of a high quality and often mistaken for American mined turquoise, the gem has never attained a vaunted status in the Chinese culture the way jade has as a precious stone.

South America was one of the earlier producers of turquoise as an ornamental and ceremonial gemstone. While few countries can attest to having major outcroppings of the gemstone, turquoise has been mined for centuries in small quantities in countries where the gem has a beautiful blue coloring. Peruvian turquoise is an excellent example of the material mined in some South American countries. Having been overshadowed by its northern neighbor, the United States, little gem grade material is associated with South America today.

United States turquoise has been the backbone of the gem cutter's mineral for over a century. Synonymous with the American Southwest, turquoise has been worked by Native Americans into some of the finest and most elaborate jewelry. It adorns those who favor the high quality and brilliant sky blue color that comes from a wonderful cabochon set in silver, free-form pieces, and those in a silver and gold intarsia piece.

Primarily located in the American Southwest, the states of Arizona, New Mexico, Nevada, and Colorado have all been hosts to many of the most famous turquoise mines in history—places where each vein yields a treasure trove of gem that is unique to its host rock. From blue to green, solid to spider web, there is no finer gemstone associated with the United States than turquoise. From legend and lore to the beautiful work of today's artisans, the turquoise of America can be found being worn by Native Americans as part of ceremonial costumes, or being sold in plazas, gift shops, and some of the finest galleries throughout the country.

Chapter 4
Turquoise Mines, Mineral, & Wearable Art

The primary intent of this volume is to focus on the various mines turquoise has been extracted from over time. Much of the text and illustrations will reflect the popular material having been extracted, cut and polished, and set from many of the most well-known and productive mines. You will however find numerous lesser-known mines discussed and illustrated with cut and polished pieces.

It is the mines of the United States that are most well known and many are highlighted through the text and images in the coming pages.

Ajax (Nevada) is one of the smaller mines in south central Nevada. In close proximity to the better-known Royston mining area, the Ajax mine is one of a relatively few newer American turquoise mines. The material mined from Ajax ranges from light blue with blue veins to an overall dark green stone with light blue coloring. [Figure 1]

Figure 1. Ajax turquoise. Necklace, multiple stone beads. Pueblo. 18" length. *$375-550.*

Bisbee (Arizona) is so named for its location to Bisbee, Arizona. Bisbee turquoise is one of the most famous in the world, and the mining of turquoise is only a part of the larger Bisbee copper mine operation. The turquoise of the Bisbee mine is recognized for its relative hardness and wonderful webbing, along with its brilliant blue coloring. Some refer to the matrix that forms the incredible natural spider web design as "Smokey Bisbee." Bisbee turquoise is found at all levels of the mine. Though the highest grade is from the 100-foot level, good quality grade is found as high as 2,000 feet. Sadly, the Phelps Dodge Mining Company, the mine's owner, recently announced Bisbee nearly depleted and the mine has been buried under tons of dirt. Bisbee will always be prized for its rarity, spider webbing, and natural beauty. [Figures 2-8]

Figure 2. Bisbee turquoise. Assorted cut and polished stone. Twenty-four grams. *Courtesy of John Renner.*

Figure 3. Bisbee II turquoise. Assorted cut and polished stone. Sixteen grams. *Courtesy of John Renner.*

Figure 4. Bisbee turquoise. Belt buckle, multiple stone. Silver setting. Navajo, signed: Justin Moris. 4" x 3". *$600-800.*

Figure 5. Bisbee turquoise and coral. Assorted belt buckles, multiple stone and chip inlay. Silver setting. Navajo. 2-1/4" x 2-1/2", 2" x 1-3/4", 2" x 1-3/4". *$200-300.*

Figure 6. Bisbee turquoise. Pendant. Silver setting. Navajo. 2-3/4" x 2-1/4". *$200-350.*

Figure 7. Bisbee turquoise. Squash blossom necklace, multiple stone. Silver setting. Navajo. 24" length. *$1300-1900.*

Figure 8. Bisbee turquoise. Cuff bracelet, multiple stone. Silver Setting. Navajo. 2-1/4" x 3-1/2". *$690-1100.*

Blue Gem (Nevada) is regarded as one of Nevada's most prolific mines for the immense variety of turquoise recovered. Found near Battle Mountain, the turquoise of Blue Gem runs from distinctively deep blues to deep greens in matrix. Blue Gem should not be confused with other Nevada mines of the same name. The Battle Mountain Blue Gem mine is now closed, though the material is still highly sought after by jewelers for its rich color. [Figures 9-18]

Figure 9. Blue Gem turquoise. Belt buckle, multiple stone. Silver setting. Navajo, signed: ew. 4-3/4" x 3". *$375-500.*

Figure 10. Blue Gem turquoise. Necklace, multiple stone, and beads. 22" length. *$1500-2000.*

Figure 11. Blue Gem turquoise. Squash blossom necklace, multiple stone. Silver setting. Navajo. 26" length. *$1200-1700.*

Figure 12. Blue Gem turquoise. Watch cuff bracelet, multiple stone inlay. Silver setting. Navajo. 1-1/4" x 3". *$400-500.*

Figure 13. Blue Gem turquoise. Cuff bracelet, multiple stone cluster. Silver setting. Navajo. 3" x 3". *$700-900.*

Figure 14. Blue Gem turquoise. Cuff bracelet. Silver setting. Signed: ES. 1" x 3". *$350-500.*

Figure 15. Blue Gem turquoise. Cuff bracelet. Silver setting. Navajo. 2" x 3". *$300-450.*

Figure 16. Blue Gem turquoise. Cuff bracelet, multiple stone. Silver setting. Navajo. 1-1/2" x 3". *$500-700.*

Figure 17. Blue Gem turquoise. Cuff bracelet. Silver setting. 2" x 3". *$400-600.*

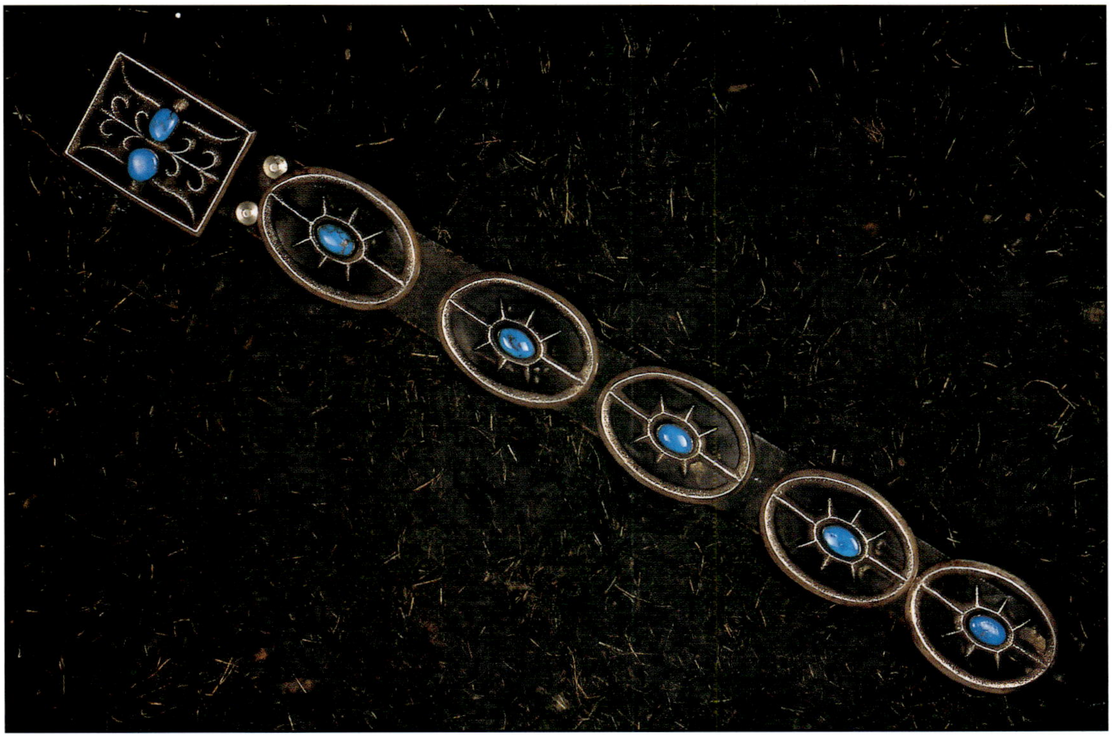

Figure 18. Blue Gem turquoise. Choncha belt. Silver setting. 2-1/2" x 2" (buckle), 2-1/4" x 1-3/4" (Choncha). *$2000-2500.*

Blue Jay (Nevada) mine is located in the Battle Mountain region of the state. It is one of the small Nevada mines to produce high-grade gem nuggets of a larger size. The blue color of this turquoise is similar to that from the Fox mine. This turquoise can also be found to have large spider webbing or brown matrix in the cut material. It is an excellent turquoise to cut and polish because of its relative hardness, and looks stunning when set in silver as a pendant or in a bracelet. [Figure 19]

Figure 19. Blue Jay turquoise. Belt buckle, multiple stone. Silver setting. Navajo. 3-1/2" x 2-1/2" length. *$450-650.*

Candelaria (Nevada) turquoise is another of the now depleted and closed Nevada mines that produced gem grade rich blue stone. Also mined from Candelaria was material that had occasional brown or black matrix. Collectors of the turquoise from the Candelaria mine know it for its beautiful quality and rarity. [Figures 20-21]

Figure 20. Candelaria turquoise. Cuff bracelet. Silver setting. Navajo. 3/4" x 3". *$200-300.*

Figure 21. Candelaria turquoise. Cuff bracelet, multiple stone. Gold setting. Signed: John Renner. 1" x 3". *Courtesy of the artist. $1500-2000.*

Carico Lake (Nevada) mine is a dry lake bed in Lander County, Nevada. Recognized for the beauty of the area, Carico Lake is at a higher elevation than many turquoise mines. While Carico Lake is most often renowned as a gold producing mine, the mining company has leased the turquoise producing area to miners and prospectors, and fine gem grade material is recovered from time to time. Because of the limited area and time to work a claim, the turquoise of this mine is a beautiful addition to any collection. You'll find some outstanding Carico Lake turquoise primarily in Irish green colors. This is due in large measure to the zinc content in the material. Even more rare is the dark blue-green Carico Lake turquoise with exceptional spider web matrix. [Figures 22-26]

Figure 22. Carico Lake turquoise. Assorted cut and polished stones. Eighteen grams. *Courtesy of John Renner.*

Figure 23. Carico Lake turquoise and coral. Necklace, multiple stone beads. Navajo. 24" length. *$350-500.*

Figure 24. Carico Lake turquoise. Bolo tie, multiple stone. Silver setting with silver tips. Navajo, signed: AZ. 3-1/2" x 2-1/2". *$800-1200.*

Figure 25. Carico Lake, Kingman, Persian, and Lone Mountain turquoise. Rings. Gold and silver settings. *$125-2200.*

Figure 26. Carico Lake turquoise. Cuff bracelet, multiple stone. Silver setting. Navajo. 1" x 3". *$250-350.*

Castle Dome (Arizona) is an open pit copper mine from which beautiful, mostly solid blue turquoise was mined. Turquoise, copper, and other minerals are often associated together in mining areas because of the nature of their chemical make-up. Castle Dome turquoise is a good example of copper mining revealing stunning stones of high-quality gem grade. [Figure 27]

Figure 27. Castle Dome turquoise. Belt buckle, multiple stone cluster. Silver Setting. Zuni, signed: W. V. Quam. 4" x 3". *$200-2500.*

Cerrillos (New Mexico) mine turquoise is not only rich in beauty, but equally rich in history. The oldest known mine of any kind in North America, Cerrillos is situated only ten miles from Santa Fe, New Mexico. With a large deposit of turquoise, some is partially exposed naturally and therefore easier to work. The early miners used stone axes, chisel and picks to pry the precious material from its host rock. It is believed that Pueblo miners who worked the mine most recovered over 100,000 tons of rock creating a mine pit some 200 feet deep. In addition, vertical shafts were also dug to reach turquoise veins. Cerrillos turquoise was used as a trade tool of these earlier people, and the Pueblos continued mining the mineral through the early 1870s. It was then that the silver boom was on, and interest in the area was heightened soon thereafter when Tiffany's of New York bought the mined area and recovered over $2,000,000 worth of turquoise in the 1890s. One of North America's most workable mined turquoise, Cerrillos is synonymous with high quality and gem grade stones. Because of its relative hardness, Cerrillos turquoise is often easier to work with and takes a more lustrous polish than the stones mined at other locations. Having been formed at the base of a volcano, Cerrillos turquoise was created by Mother Nature under more unusual circumstances than most turquoise. This occurrence, the only specific known type of its kind, has given Cerrillos a plethora of colors created from other minerals in the numerous host rocks. Over seventy-five various color shades have been identified to date. These range from the rather bland yet rare light brown or tan to extremely vibrant blue-green gem stones. [Figures 28 - 36]

Figure 28. Cerrillos turquoise. Assorted cut and polished stones. Fifty-six grams. *Courtesy of John Renner.*

Figure 29. Cerrillos turquoise. Pins, multiple stone. Silver setting. Navajo. 2-3/4" x 1-1/4", 2-1/4" x 2". *$200-300.*

Figure 30. Cerrillos turquoise. Pin. Silver setting. Navajo. 1-3/4" x 1/2". *$175-275.*

Figure 31. Cerrillos turquoise and coral. Cuff bracelet, multiple stone. Silver setting. Navajo. 1" x 3". *$1500-2000.*

Figure 32. Cerrillos turquoise. Cuff bracelet. Silver setting. Navajo. 1" x 3". *$450-600.*

Figure 33. Cerrillos turquoise. Cuff bracelet. Silver setting. 2" x 3". *$400-550.*

Figure 34. Cerrillos turquoise. Cuff bracelet. Silver setting. 2" x 3". *$350-475.*

Figure 35. Cerrillos turquoise. Cuff bracelet, multiple stone. Silver setting. Signed: John Renner. 2" x 3-1/2". *$600-800. Courtesy of the artist.*

Figure 36. Cerrillos turquoise. Cuff bracelet, multiple stone. Silver setting. Signed: John Renner. 1-1/4" x 3". *$1100-1500.* *Courtesy of John Renner.*

Chinese (China) turquoise is the predominant material on today's market. Accounting for eighty percent of all turquoise in the United States market, nearly all Chinese turquoise has been stabilized to harden the stone before being set into jewelry. The majority of Chinese turquoise being mined comes from the Hubei Province that produces turquoise very similar to that of depleted Southwestern United States mines, particularly those in Nevada—hence, the confusion that can be caused for collectors and jewelers. Other deposits have been worked from the Ma'ashan mine, northwest of the large city of Shangahai. In the Academy of Social Sciences in Bejing there is a bronze with turquoise overlay plaque dated to the Erlitou site giving rise to the belief that turquoise mining dates to 1700 B.C.E. in China. Most early turquoise in the region is credited to trade between the Chinese, Persians, Turks, and others. More specific information on the mining operations today is not as readily available as some other countries, due in part to governmental controls. [Figures 37-97]

Figure 37. Chinese turquoise. Assorted cut and polished stones. Forty-six grams total. *Courtesy of John Renner.*

Figure 38. Chinese turquoise. Belt buckle. Silver setting. Navajo, signed: H.B. 3-1/2" x 3". *$350-450.*

Figure 39. Chinese turquoise. Belt buckle. Silver setting. Navajo, signed: VY. 4-1/2" x 3". *$400-600.*

Figure 40. Chinese turquoise. Belt buckles. Silver settings. Navajo. 3" x 2-1/2". *$400-600.*

Figure 41. Chinese turquoise. Belt buckle. Silver setting. Navajo, signed: H.B. 3-1/2" x 3". *$350-450.*

Figure 42. Chinese turquoise. Choncha belt, multiple stone. Silver setting. 3" x 2-1/2", 2-1/2" x 2". *$2000-4000.*

Figure 43. Chinese turquoise. Necklace, beads. Navajo. 24" length. *$350-450.*

Figure 44. Chinese turquoise. Necklace, multiple stone, double strand beads. Navajo. 24" length. *$250-350.*

Figure 45. Chinese turquoise and coral. Necklace, multiple stone, eight strand. Navajo. 22" length. *$400-600.*

Figure 46. Chinese turquoise; assorted minerals. Necklace, multiple stone, three strand. Pueblo. 24" length. *$375-550.*

Figure 47. Chinese turquoise; coral and shell. Necklace, multiple stone, double strand beads. Navajo. 20" length. *$200-300.*

Figure 48. Chinese turquoise. Necklace, multiple stone nuggets. Navajo. 12" length. *$175-275.*

Figure 49. Chinese turquoise and coral. Necklace, multiple stone, five strands. Navajo. 20" length. *$350-450.*

Figure 50. Chinese turquoise. Pin. Silver and brass setting, multiple stone. Navajo. 2-1/2". *$65-150.*

Figure 52. Chinese turquoise. Pin. Silver setting. 1-1/2" x 1-1/2". *$100-150.*

Figure 51. Chinese turquoise. Pin. Silver setting. Navajo. 2-1/2" x 2". *$175-300.*

Figure 53. Chinese turquoise. Pendant. Silver setting. Navajo, signed: Santa Fe DEU. 2" x 1-3/4". *$100-150.*

Figure 54. Chinese turquoise. Pendant, multiple stone. Silver setting. Navajo. 2" x 1". *$100-150.*

Figure 56. Chinese turquoise. Pendant. Silver setting. 2-1/4" x 2". *$125-200.*

Figure 55. Chinese turquoise. Pendant, multiple stone cluster. Silver setting. 2" x 1". *$125-175.*

Figure 57. Chinese turquoise. Pendant. Silver setting. Navajo, signed: H. 2-1/2" x 2". *$50-100.*

Figure 58. Chinese turquoise. Pendant. Silver setting. Navajo. 1-3/4" x 1-1/2". *$75-150.*

Figure 61. Chinese turquoise. Pendant. Silver setting. Navajo, signed: E. 2-1/2" x 1". *$75-150.*

Figure 59. Chinese turquoise. Pendant. Silver setting. Navajo, signed: DB. 2-1/2" x 1-1/4". *$75-150.*

Figure 62. Chinese turquoise. Pendant. Silver setting. Navajo, signed: CM. 2-1/2" x 3-1/4". *$200-300.*

Figure 60. Chinese turquoise. Pendants. Silver settings. Navajo, signed: DB. 1" x 3/4", 1-1/4" x 7/8", 1" x 1". *$50-150.*

Figure 63. Chinese turquoise. Pendant. Silver setting. Navajo, signed: D. 3-1/4" x 1-3/4". *$200-300.*

Figure 65. Chinese turquoise. Pendant. Silver setting. Navajo, signed: AS. 2-1/2" x 1". *$50-100.*

Figure 64. Chinese turquoise, coral and lapis. Pendant, multiple stone and inlay. Silver setting. Navajo, signed: L. 2" x 1". *$75-150.*

Figure 66. Chinese turquoise. Pendant. Silver setting. Navajo, signed: DB. 2" x 1". *$50-100.*

Figure 68. Chinese turquoise. Pendant. Silver setting. Navajo, signed: DB. 3-1/2" x 1-1/4". *$150-225.*

Figure 67. Chinese turquoise. Pendant. Silver setting. 1-3/4" x 3/4". *$50-100.*

Figure 69. Chinese turquoise. Pendant. Silver setting. Navajo, signed: E. 2" x 1". *$90-125.*

Figure 70. Chinese turquoise. Pendant. Silver setting. Signed: John Renner. 4-3/4" x 3". *Courtesy of the artist. $550-750.*

Figure 71. Chinese turquoise. Pendant. Silver setting, liquid silver strands. Navajo, signed: Elouise. 12" length. *$375-500.*

Figure 72. Chinese turquoise. Pendant, beads. Silver setting. Navajo, signed: DB. 20" length. *$125-175.*

Figure 73. Chinese turquoise. Pendant, beads. Silver setting. Navajo, signed: DB. 20" length. *$125-175.*

Figure 74. Chinese turquoise. Pendant. Silver setting. Navajo, signed: E. Etsity. 4-3/4" x 3-1/4". *$500-700.*

Figure 75. Chinese turquoise. Necklace, multiple stones. Silver setting. Navajo. 28" length. *$800-1000.*

Figure 76. Chinese turquoise and coral. Necklace, beads. 20" length. *$200-300.*

Figure 77. Chinese turquoise. Belt buckle. Silver setting. Signed: Brad Davis. 4" x 4". *$200-350.*

Figure 78. Chinese turquoise. Pin. Silver setting. Navajo. 3" x 2-3/4". *$350-500.*

Figure 79. Chinese turquoise. Earrings, cluster. Silver setting. Navajo. 1-1/2". *$100-150.*

Figure 80. Chinese turquoise. Cuff bracelet, multiple stone cluster. Silver setting. 2-1/2" x 3". *$600-800.*

Figure 81. Chinese turquoise. Cuff bracelet. Silver setting. Navajo. 3" x 3". *$300-400.*

Figure 82. Chinese turquoise. Cuff bracelet. Silver setting. Navajo. 2-1/2" x 2". *$100-200.*

Figure 83. Chinese turquoise. Cuff bracelet. Silver setting. Navajo. 4" x 3-1/2". *$350-450.*

Figure 84. Chinese turquoise, coral, lapis, and black onyx. Cuff bracelet. Silver setting. Navajo. 2-1/2" x 3". *$275-400.*

Figure 85. Chinese turquoise. Cuff bracelet. Silver setting. Navajo. 3-1/2" x 3". *$200-300.*

Figure 86. Chinese turquoise and coral. Cuff bracelet, multiple stone cluster. Silver setting. Navajo, signed: Kirt Smith. 3-1/2" x 3". *$450-650.*

Figure 87. Chinese turquoise. Cuff bracelet, multiple stone. Silver setting. Signed: Nakai. 1" x 3". *$350-500.*

Figure 88. Chinese turquoise. Cuff bracelet. Silver setting, multiple stone. Navajo, signed: DM. 1-1/4" x 3". *$300-400.*

Figure 89. Chinese turquoise. Cuff bracelet, multiple stone inlay. Silver setting. Navajo, signed: Lon Parker. 3/4" x 3". *$500-700*.

Figure 90. Chinese turquoise. Cuff bracelet, multiple stone. Silver setting. Signed: Tom Dewit. 1" x 3". *$600-900.*

Figure 91. Chinese turquoise, coral and shell. Cuff bracelet, multiple stone inlay. Silver setting. 2" x 3". *$900-1500.*

Figure 92. Chinese turquoise. Cuff bracelet, multiple stone. Silver setting. Navajo, signed: ES. 3/4" x 3". *$500-700.*

Figure 93. Chinese turquoise. Cuff bracelet. Silver setting. Navajo, signed: Padilla. 2-3/4" x 3". *$400-600.*

Figure 94. Chinese turquoise, assorted gems. Pins. Gold settings. Signed: E.S. 3-1/2" x 3-1/2". *N/A*

Figure 95. Chinese turquoise, assorted gems. Pin. Silver setting. Signed: E.S. 3-1/2" x 3-1/2". *N/A*

Figure 96. Chinese turquoise, assorted gems. Pins. Silver settings. Signed: E.S. 3" x 3-1/4". *N/A*

Figure 97. Chinese turquoise, assorted gems. Pins. Silver settings. Signed: E.S. 3-1/2" x 3-1/2". *N/A*

Cripple Creek (Colorado) in Colorado is known primarily as a gold producing site. There are a number of small mines in the region where turquoise has been found as a by-product of the gold mining operations. Cripple Creek turquoise is usually green in color, though there has been some blue with matrix material found as well. A rare addition to any collection of turquoise is cut and polished stone from this area. [Figure 98]

Figure 98. Cripple Creek turquoise. Watch cuff bracelet. Silver setting. Navajo. 1-1/2" x 3". *$450-600.*

Figure 99. Damele turquoise. Cuff bracelet, multiple stone. Silver setting. Signed: John Renner. 3/4" x 3". *$650-850. Courtesy of the artist.*

Damele (Nevada) turquoise is found close to the Carico Lake mine area. The distinctive material is found as a hard yellow-green stone largely due to the higher than normal zinc content. It is most often found with spider webbing of dark brown or black and is limited because of the small size of the mine itself. One of the rare non-blue colors of turquoise, Damele is collectible for its yellowish color. [Figures 99-101]

Figure 100. Damele turquoise. Choncha belt, multiple stone. Silver setting. Navajo. 3-3/4" x 3-1/4". *$1800-2400.*

Figure 101. Damele variscite. Assorted cut and polished stone. Six grams. *Courtesy of John Renner.*

Dry Creek (Nevada) has been known by a few names, including Godber and Burnham mine. It is one of the Nevada mines that brought forth turquoise of creamy white to pale blue. Well-known to collectors and specialists for its rare white coloring, the turquoise is quite hard and its color is caused by an unusually high amount of aluminum in relation to the copper content in the stone. The matrix is most often a pleasing light yellow, gray or black color. This turquoise is highly collectible, though you will find it set in jewelry most often with other more traditional blue colored stones. [Figure 102]

Figure 102. Dry Creek turquoise. Cuff bracelet. Silver setting. Signed: John Renner. 1-1/2" x 3". *Courtesy of John Renner. $650-850.*

Easter Blue (Nevada) mined turquoise is some of the oldest to come out of Nevada. The mine is situated in the Royston area and has produced gem grade blue-green with matrix material. A classic example turquoise from this region, Easter Blue sets well alone or with other stones. [Figures 103-106]

Figure 103. Easter Blue turquoise. Assorted cut and polished stones. Fifty-four grams. *Courtesy of John Renner.*

Figure 104. Easter Blue turquoise and coral. Belt buckle, multiple stone. Navajo. 5" x 3-1/2". *$650-750.*

Figure 105. Easter Blue turquoise. Cuff bracelet, multiple stone. Silver setting. Navajo. 2-1/2" x 3". *$450-650. (View one)*

Figure 106. Easter Blue turquoise. Cuff bracelet, multiple stone. Silver setting. Navajo. 2-1/2" x 3". *$450-650. (View two)*

Emerald Valley (Nevada) turquoise is much as its name implies, a deep and beautiful lush green material. The mine is still worked today and the stone is usually heavily mixed with matrix of light tan to brown. A unique turquoise like Emerald Valley is outstanding to have cut and polished for jewelry. [Figure 107]

Figure 107. Emerald Valley turquoise. Pendant. Silver setting. Navajo, signed: E. 2-1/4" x 1-1/4". *$75-150.*

Fox (Nevada) mine turquoise holds the distinction of being the largest produced turquoise mine in Nevada. Located near Lander County, this turquoise was first mined in prehistoric times by early peoples. Some might recognize this mine from another name it held at one time, the Cortez. Rediscovered around 1911, the Fox mine has brought forth over a half million pounds of material. While most recognize the Fox name, it is rightfully called White Horse Fox for the region in Lander County where this high-grade blue gem has been mined. [Figures 108-114]

Figure 108. Fox turquoise. Assorted cut and polished stone. Twenty-two grams. *Courtesy of John Renner.*

Figure 109. Fox turquoise. Bolo tie. Silver setting with silver tips. Navajo, signed: DM. 2-1/4" x 2". *$475-600.*

Figure 110. Fox turquoise. Cuff bracelet, multiple stone. Silver setting. Navajo. 1" x 3". *$200-300.*

Figure 111. Fox turquoise. Cuff bracelet, multiple stone. Silver setting. Navajo. 2" x 3". *$450-650.*

Figure 112. Fox (Wild Horse) turquoise. Cuff bracelet. Silver setting. Navajo, signed: Carl Quintano. 1-3/4" x 3". *$250-400.*

Figure 113. Fox (Wild Horse) turquoise. Various pendants. Silver setting. Navajo. left, middle, right. 1-1/4" x 1", 2" x 1", 1" x 1". *$100-300.*

Figure 114. Fox (Wild Horse) turquoise. Choncha belt, multiple stone in cluster setting. Signed: C. L. 3-1/2" x 2-3/4". *$1,800-2,400.*

Godber and Burnham (Nevada) mine turquoise is also referred to as Dry Creek turquoise. The mine is located in northeast Nevada and most often this stone is found as white to light blue. Turquoise from this locale is quite hard due to the presence of more aluminum in its make up. The material is nearly always with matrix that is light yellow or brown/gray to gray/black. [Figures 115-116]

Figure 115. Godber Burnham turquoise. Belt buckle, multiple stone. Silver setting. Navajo, signed: M. John. 5" x 2-1/2". *$500-700.*

Figure 116. Godber Burnham turquoise. Pendant. Silver setting. 1-3/4" x 1-1/2". *$200-300.*

Hachita (New Mexico) turquoise has not been mined since 1905. While it dates back to ancient times, it is a rare and difficult to acquire turquoise. It is generally a deep blue material and can be confused with the turquoise of Tyrone. [Figures 117-124]

Figure 117. Hachita and Chinese turquoise. Belt buckle, multiple stone, cluster. Silver setting. Navajo, signed: DC. 3-1/2" x 2-1/4". *$350-475.*

Figure 118. Hachita turquoise. Belt buckle. Silver setting. Navajo, signed: WBH. 2-1/2" x 2". *$200-350.*

Figure 119. Hachita turquoise. Belt buckle. Silver setting. Navajo. 3" x 2-1/2". *$100-200.*

Figure 120. Hachita turquoise. Pin. Silver setting. Navajo, signed: Santa Fe DEU. 3" x 1-1/2". *$80-150.*

Figure 121. Hachita turquoise. Pendant, multiple stone. Silver setting. Navajo, signed: Ray Begay. 3-1/2" x 1-1/4". *$150-225.*

Figure 122. Hachita turquoise. Cuff bracelet. Silver setting. Navajo. 3" x 3-1/2". *$275-425.*

Figure 123. Hachita turquoise. Choncha belt, multiple-stone cluster. Silver setting. Navajo. 3-1/2" x 2-1/2", 2-1/4" x 1-1/2", 2-1/4". *$1750-2250.*

Figure 124. Hachita turquoise. Choncha belt, multiple stone. Silver setting. Navajo. 3-1/2" x 3", 3" x 2-3/4". *$2000-2500.*

Indian Mountain (Arizona) turquoise was discovered in 1970 by a Native American Shoshone sheepherder. Recognized as the best known of all contemporary turquoise mines, the vein of turquoise was subsequently mined by an American family and their marketing efforts were geared to some of the most well-known Southwestern jewelry artists. The beautiful blue of Indian Mountain lends itself to some of the finest pieces of modern Native American wearable art. In the late 1970s, turquoise from this rich lode was prominently featured in the highly regarded Western magazine, *Arizona Highways*. The issue of the state magazine remains an important collectible for anyone in the turquoise collecting field. [Figures 125-128]

Figure 125. Indian Mountain turquoise. Cuff bracelet, multiple stone. Silver setting. Navajo. 1-3/4" x 3". *$500-700.*

Figure 126. Indian Mountain turquoise. Cuff bracelet, multiple stone. Silver setting. Signed. 1" x 3". *$500-700.*

Figure 127. Indian Mountain turquoise. Cuff bracelet. Silver setting. Signed: John Renner. 3-3/4" x 3". *Courtesy of the artist. $800-1100.*

Figure 128. Indian Mountain turquoise. Choncha belt. Silver setting. Signed: D Gordon. 4-1/2" x 3-1/4"; 3-1/4", *$2000-2500.*

Kingman (Arizona) mined turquoise has become a standard by which many other turquoise stones are judged. One of the largest turquoise mines in North America, Kingman has a stature all its own—a stature that is reflected in the term "Kingman," referring to the intense blue coloring this gemstone most often exhibits. While few mines produced turquoise nuggets, Kingman natural nuggets are famed for their roundness and matrix. One of the most recognized names in turquoise the world over, Kingman is truly a leader in high quality cut and polished stones. [Figures 129-239]

Figure 129. Kingman turquoise. Natural rough stone. Assorted stones. 220 grams. *Courtesy of John Renner.*

Figure 130. Kingman turquoise. Assorted natural rough, cut and polished stones. Fourteen grams. *Courtesy of John Renner.*

Figure 131. Kingman turquoise. Polished stone. *Courtesy of John Renner.*

Figure 132. Kingman turquoise. Belt buckle, multiple stone cluster. Navajo, signed JR. 3-1/4" x 3". *$250-350.*

Figure 133. Kingman turquoise. Belt buckle. Signed: John Renner. 4" x 2-1/2", right; 4" x 6", left. *Courtesy of the artist. $300-1000.*

Figure 134. Kingman turquoise. Belt buckle. Signed: John Renner. 5" x 4-1/2". *Courtesy of the artist. $800-1000.*

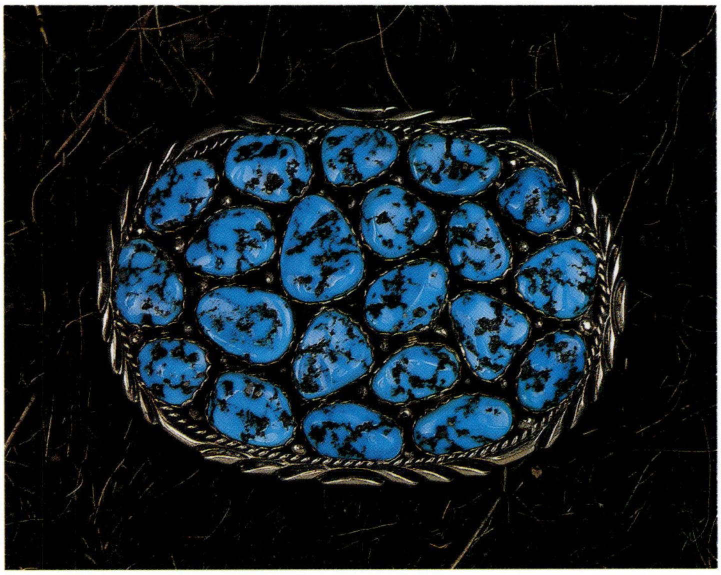

Figure 135. Kingman turquoise. Belt buckle, multiple stone cluster. Navajo, signed: Yazzie. 2-1/2" x 3-1/2". *$450-600.*

Figure 136. Kingman turquoise. Belt buckle, multiple stone. Signed: John Renner. 3" x 5". *Courtesy of the artist. $500-750.*

Figure 137. Kingman turquoise. Belt buckle, multiple stone cluster. Navajo, signed: Tommy Moore. 3-1/4" x 4-1/4". *$400-600.*

Figure 138. Kingman turquoise. Belt buckle. Navajo. 4-1/4" x 3-3/4". *$375-500.*

Figure 139. Kingman turquoise. Belt buckle. Navajo. 2-1/2" x 3-1/2". *$100-200.*

Figure 140. Kingman turquoise. Belt buckle. Signed: John Renner. 4-1/2" x 6". *Courtesy of the artist.* *$750-1000.*

Figure 141. Kingman turquoise, coral, mother-of-pearl and malachite. Belt buckle, multiple stone and inlay. Navajo. 2" x 3". *$100-200.*

Figure 142. Kingman turquoise. Belt buckle, multiple stone cluster. Silver setting. Navajo. 2-1/2" x 3". *$250-400.*

Figure 143. Kingman turquoise. Belt buckle, multiple stone. Signed: John Renner. 2-1/4" x 3-1/4". *Courtesy of the artist. $275-400.*

Figure 144. Kingman turquoise. Belt buckle, multiple stone and inlay. Silver setting. 4-1/4" x 5-3/4". *$450-650.*

Figure 145. Kingman turquoise. Belt Buckle. Navajo. 2-3/4" x 3-1/4". *$200-300.*

Figure 146. Kingman turquoise. Necklace, multiple stone with silver beads. Navajo. 12" length. *$350-500.*

Figure 147. Kingman turquoise. Necklace, multiple stone with silver beads. Navajo. 24" length. *$150-300.*

Figure 148. Kingman turquoise. Necklace, multiple stone. Navajo. 30" length. *$450-600.*

Figure 149. Kingman turquoise. Necklace, multiple stone. Pueblo. 22" length. *$300-450.*

Figure 150. Kingman turquoise. Necklace, multiple stone with silver. Pueblo. 28" length. *$450-600.*

Figure 151. Kingman turquoise. Pendant, multiple stone cluster. Navajo, signed: V. 4-1/4" x 6". *$600-800.*

Figure 152. Kingman turquoise and coral. Pendants, multiple stone. Navajo, signed: JR (Jerry Roan). 4" x 6". *$800-1100.*

Figure 153. Kingman and Chinese turquoise and coral. Necklaces, multiple stone, multi-strand. Navajo. 22", 28" length. *$200-800.*

Figure 154. Kingman and Block turquoise, multiple stone, four-strand. Kingman turquoise with coral, top; Block turquoise, bottom. Necklace. Navajo. 30". *$200-400.*

Figure 155. Kingman and Chinese turquoise and coral. Cross pendants and pin, multiple stone. Silver setting. Navajo. 1" - 2-1/2" length. *$100-400.*

Figure 156. Kingman turquoise, agate and onyx. Pendants, multiple stone inlay. Silver setting. Pueblo. 2". *$100-200.*

Figure 158. Kingman turquoise. Pendants, multiple stone. Silver setting. Navajo. 1" x 3", 2". *$100-275.*

Figure 157. Kingman turquoise. Pin. Silver setting. Navajo. 2". *$150-250.*

Figure 159. Kingman turquoise, coral, shell, onyx and agate. Pin, multiple stone, inlay. Silver setting. 1" x 3". *$200-400.*

Figure 160. Kingman turquoise. Necklace, multiple stone inlay, four-strand beads. Silver setting. Pueblo. 22" length. *$450-650.*

Figure 161. Kingman turquoise and coral. Fetish necklace, multiple stones. Pueblo. 20". *$900-1100.*

Figure 162. Kingman turquoise. Necklace, multiple stone beads with silver. Navajo. 18". *$1100-1500.*

Figure 164. Kingman turquoise. Pendant, multiple stone. Silver setting. Navajo, signed: Toadlee. 2-3/4" x 3". *$200-400.*

Figure 163. Kingman turquoise. Pendant. Silver setting. Navajo, signed: LLY. 2". *$150-300.*

Figure 165. Kingman turquoise. Pendant. Silver setting. 2". *$100-200.*

Figure 166. Kingman turquoise. Pendant. Silver setting. Navajo, signed: E. 1-1/2" x 2-3/4". *$100-200.*

Figure 167. Kingman turquoise. Pendant. Silver setting. Navajo, signed: E. 1-1/2". *$100-200.*

Figure 168. Kingman turquoise. Pendant, multiple stone. Silver setting. Navajo. 3-1/2" x 1". *$150-300.*

Figure 169. Kingman turquoise. Pendant. Silver setting. Navajo, signed: E. 2". *$100-200.*

Figure 170. Kingman turquoise. Pendant. Silver setting. Navajo, signed: E. 1-1/2" x 3-1/4'. *$200-350.*

Figure 171. Kingman turquoise and coral. Bear claw pendants, multiple stones and silver. Navajo. 1" - 3". *$200-350.*

Figure 172. Kingman turquoise Shell and Cerrillos turquoise, multiple stones. Necklace. Pueblo. 14" length. *$500-1000.*

Figure 173. Kingman turquoise. Pendant, multiple stones. Silver setting. Signed: John Renner. 18" length. *Courtesy of the artist. $600-900.*

Figure 174. Kingman turquoise. Pendant. Silver setting. Bead necklace. Navajo, signed: E. 18" length. *$150-300.*

Figure 175. Kingman turquoise. Necklace, multiple stones. Silver setting. Signed: John Renner. 18" length. *Courtesy of the artist. $600-900.*

Figure 176. Kingman turquoise. Necklace and earrings, multiple stone cluster. Silver setting. Navajo, signed: ZB. 20" length. *$1500-2250.*

Figure 177. Kingman turquoise. Pendant, multiple stones. Silver setting. Signed: John Renner. 18" length. *Courtesy of the artist. $550-750.*

Figure 178. Kingman turquoise. Pendant. Silver setting. Navajo, signed: C. 2-3/4" x 3-3/4" length. *$100-200.*

Figure 179. Kingman and Chinese turquoise. Necklace. Shell: Kingman turquoise, inlay; beads: Chinese turquoise, six strands. Navajo, signed: C. 2-1/4" x 2-1/2" shell, 18" length. *$1500-2000.*

Figure 180. Kingman turquoise, coral, onyx and mother-of-pearl. Necklace, multiple stone inlay. Silver setting. Zuni. 18" length. *$1100-1500.*

Figure 181. Kingman turquoise. Pendant and earring set, multiple stones. Silver setting. Signed: DB. Pendant: 1-1/4" x 2-1/4", earrings: 1-1/4" x 1/2". *$250-350.*

Figure 182. Kingman turquoise (Reconstituted). Fetish necklace. Navajo. 18" length. *$150-250.*

Figure 183. Kingman turquoise (Reconstituted), agate, mother-of-pearl. Fetish necklace. Navajo. 18" length. *$150-300.*

Figure 184. Kingman and Chinese turquoise. Pendant: Kingman; necklace Chinese, four-strand beads: Chinese. Navajo. 18" length. *$600-800.*

Figure 185. Kingman Turquoise. Necklace, multiple stones, silver with silver beads. Navajo. 18" length. *$600-900.*

Figure 186. Kingman turquoise and coral. Belt buckles, chip inlay. Silver settings. Navajo, signed: S, (left), Denesto, (middle), HB, (right). 3" x 2". *$250-650.*

Figure 187. Kingman turquoise. Belt buckle, multiple stones. Silver setting. Signed: John Renner. 4" x 3-1/2". *Courtesy of the artist.* *$450-700.*

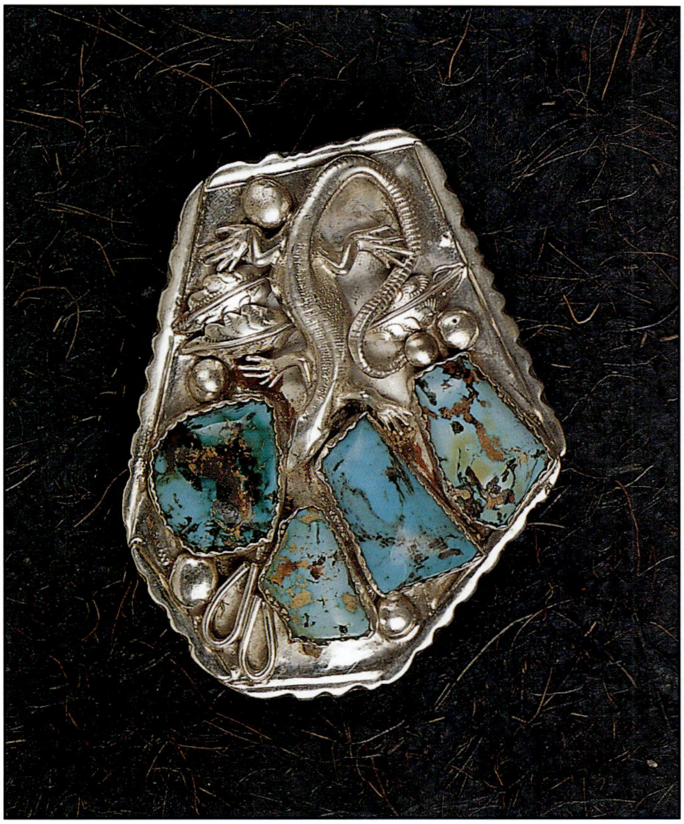

Figure 188. Kingman turquoise. Belt buckle, multiple stones. Silver setting. Signed: John Renner. 4" x 3-1/2". *Courtesy of the artist.* *$450-700.*

Figure 189. Kingman turquoise. Earrings, multiple stones. Silver setting. Zuni. 2-1/4". *$150-250.*

Figure 190. Kingman turquoise. Earrings, multiple stone cluster. Silver setting. Navajo. 2" x 2-1/4". *$150-250.*

Figure 191. Kingman turquoise. Earrings. Silver setting. Navajo, signed: GD. 7/8". *$50-150.*

Figure 192. Kingman turquoise. Earrings. Silver. 5/8". *$25-50.*

Figure 193. Kingman turquoise. Earrings, multiple stones. Silver bear claw setting. Navajo. 1-1/4" x 2-1/2". *$25-50.*

Figure 194. Kingman turquoise. Earrings, multiple stones. Silver. Navajo. 1-1/4". *$25-50.*

Figure 195. Kingman turquoise. Earrings, multiple stones. Silver setting. Navajo. 1/4" x 1-1/2". *$30-60.*

Figure 196. Kingman turquoise. Earrings, multiple stones. Silver setting. Navajo. 1-1/4". *$25-50.*

Figure 197. Kingman turquoise. Bolo tie. Silver setting and tips. Signed: John Renner. 2-1/4" x 3". *Courtesy of the artist. $175-275.*

Figure 198. Kingman turquoise. Bolo tie, multiple stones. Silver setting with silver tips. Navajo, signed: JR. 6" x 4". *$800-1100.*

Figure 199. Kingman turquoise and coral. Watch bands, multiple stone chip inlay. Silver settings, stretch bands. Navajo. 4", 5". *$75-150.*

Figure 200. Kingman turquoise. Watch band, multiple stones. Silver settings, stretch bands. Navajo. 5". *$100-150.*

Figures 201-202. Kingman turquoise. Watch cuff bracelet, multiple stones. Silver setting. Navajo, signed: PB. 3" w. *$350-450.*

Figure 203. Kingman turquoise. Watch cuff bracelet, multiple stones. Silver setting. Navajo. 1-1/2" w. *$200-300.*

Figures 204-205. Kingman turquoise. Watch cuff bracelet, multiple stones. Silver setting. Navajo. 2-1/2" w. *$300-400.*

Figure 206. Kingman turquoise. Watch cuff bracelet, chip inlay. Silver setting. Navajo, signed: T. 2-1/2" w. *$250-350.*

Figure 207. Kingman turquoise. Cuff bracelet. Gold setting. Signed: John Renner. 2-1/2" x 3". *Courtesy of the artist. Photo: Courtesy of John Renner. $3000-4500.*

Figure 208. Kingman turquoise and faceted topaz. Cuff bracelet. Gold setting. Signed: John Renner. 1-1/2" x 3". *Courtesy of the artist. Photo: Courtesy of John Renner. $5000-7000.*

Figure 209. Kingman turquoise and coral. Watch cuff bracelet, multiple stones. Silver setting. Navajo. 2-1/2" w. *$375-500.*

Figures 210-211. Kingman turquoise. Cuff bracelet, multiple stone cluster. Silver setting. Navajo, signed: T. 4" x 3-1/2" w. *$800-1100.*

Figures 212-213. Kingman turquoise. Cuff bracelet, multiple stone cluster. Silver setting. Navajo, signed: T. 2-1/2" x 3-1/2" w. *$600-800.*

Figure 214. #8, Kingman and Pilot Mountain turquoise. Rings. Gold and silver settings. *$200-1400.*

Figure 215. Kingman turquoise and coral. Cuff bracelet, multiple stone chip inlay. Silver setting. Navajo, signed: T. 2-1/4" x 3-1/2" w. *$600-800.*

Figure 216. Kingman turquoise and coral. Cuff bracelet, multiple stone chip inlay. Silver setting. Navajo. 3" x 3" w. *$300-450.*

Figure 217. Kingman turquoise. Cuff bracelet, multiple stone cluster. Silver setting. Navajo. 2-1/4" x 3" w. *$275-425.*

Figures 218-219. Kingman turquoise. Cuff bracelet, multiple stone cluster. Silver setting. Navajo. 3" x 3" w. *$900-1200.*

Figure 220. Kingman turquoise. Bracelet, multiple stone nugget cluster. Silver setting. Navajo. 7/8" x 3" w. *$450-650.*

Figure 221. Kingman turquoise. Cuff bracelet, multiple stone cluster. Silver setting. Navajo. 4" x 3-1/2" w. *$450-650.*

Figure 222. Kingman turquoise. Bracelet, multiple stones. Silver setting. Signed: John Renner. 2-1/5" x 3" w. *Courtesy of the artist. $300-500.*

Figure 223. Kingman turquoise and coral. Cuff bracelet, multiple stones. Silver setting. Navajo. 2-1/2" x 3-1/2" w. *$500-700.*

Figure 224. Kingman turquoise. Cuff bracelet, multiple stones. Silver setting. Signed: John Renner. 3" x 3-1/2" w. *Courtesy of the artist. $700-900.*

Figure 225. Kingman turquoise. Cuff bracelet, multiple stones. Silver setting. Signed: John Renner. 1-1/2" x 3" w. *Courtesy of the artist. $500-700.*

Figure 226. Kingman turquoise. Cuff bracelet, multiple stone nugget. Silver setting. Signed: John Renner. 1-1/2" x 3" w. *Courtesy of the artist. $300-500.*

Figure 227. Kingman turquoise. Cuff bracelet, multiple stones. Silver setting. Signed: John Renner. 2" x 3" w. *Courtesy of the artist. $300-500.*

Figure 228. Kingman turquoise. Cuff bracelet, multiple stones. Silver setting. Navajo. 2-1/2" x 3". *$250-450.*

Figure 229. Kingman turquoise, agate, and coral. Assorted animal carvings. 1" - 4" length. *$30-100.*

Figure 230. Kingman turquoise. Bracelet. Silver setting. Signed: John Renner. 4" x 3". *Courtesy of the artist. $800-1000.*

Figure 231. Kingman turquoise. Cuff bracelet, multiple stones. Silver setting. Navajo. 2-1/2" x 3". *$500-700.*

Figure 232. Kingman turquoise. Cuff bracelet, multiple stone cluster. Silver setting. Navajo. 4" x 3-1/2". *$450-600.*

Figure 233. Kingman turquoise. Choncha belt, multiple stones. Silver setting. Signed: John Renner. 4-3/4" x 4-1/4", buckle; 4-1/2" x 4-3/4", Choncha. *Courtesy of the artist. $7000-9000.*

Figure 234. Kingman turquoise. Choncha belt, multiple stones. Silver setting. Signed: John Renner. 4-3/4" x 4-1/4", buckle; 4-1/2" x 4-3/4", Choncha. *Courtesy of the artist. $7000-9000.*

Figure 235. Kingman turquoise. Choncha belt, multiple stones. Silver setting, leather. 2-1/4" x 50". *$2200-2700.*

Figure 236. Kingman turquoise. Choncha belt, multiple stone cluster. Silver setting, leather. 2-1/4" x 50". *$2200-2700.*

Figure 237. Kingman turquoise. Belt, multiple stones. Silver setting, leather. Signed: John Renner. 2" x 50". *$2100-2500.*

Figure 238. Kingman and Chinese turquoise, coral, lapis. Rings. Silver settings. Kingman turquoise, coral, and lapis, left; Kingman turquoise, coral and lapis, middle; Chinese turquoise, right. *$700-900.*

Figure 239. Kingman turquoise. Necklace, multiple stones and natural rough. Gold setting. Signed: John Renner. 1-3/4" x 1-1/4" stones. *Courtesy of the artist. $1500-2000.*

Lander Blue (Nevada) mined turquoise has a uniquely Nevada experience associated with the way it was discovered. Part of this discovery leads to the value of the turquoise itself. In 1973, Rita Hapgold, a Nevada blackjack dealer, collected some turquoise nuggets while picnicking at Indian Creek. She went on to claim the site, naming it the Mary Louise Lode Mining Claim. Later that same year she sold her claim to two others who established the Lander Blue Turquoise Company. Lander Blue is recognized to be the most valuable pound for pound turquoise in the world. Referred to by many as a "hat mine" because a hat could cover the recovery area, Lander Blue gave up a meager 98 pounds of turquoise, about two bucketfuls. Not only rich in contemporary history, but exquisite in the rich dark blue with dark matrix, Lander Blue is of the highest grade turquoise and regarded as a Southwestern classic. Whether set in jewelry or as a stand-alone cut and polished stone, Lander Blue is likely to be the most expensive ounce per ounce turquoise in any collection or wearable art. [Figures 240-243]

Figure 240. Landers turquoise. Cuff bracelet. Silver setting. Navajo. 2-1/2" x 3". *$1700-2100.*

Figure 241. New Landers turquoise. Cuff bracelet, multiple stones. Silver setting. Navajo. 1-3/4" x 3". *$400-600.*

Figure 242. New Landers turquoise. Earrings, multiple stones. Silver setting. Navajo. 1-1/4" x 1/2". *$175-300.*

Figure 243. Landers, Sleeping Beauty, and Kingman turquoise. Rings. Silver settings. *$150-1500.*

Lone Mountain (Nevada), also known as **Blue Jay Mine,** is one of those enchanting Nevada turquoise mines for which the gem material is known for maintaining its color over time. Located in Esmeralda County, the turquoise is most often found in nodules and has a fairly wide blue color range, some including spider webbing. Lapidaries and jewelers have a particular fondness for Lone Mountain because it holds its color so well and therefore allows for spectacular wearable jewelry art. [Figures 244-247]

Figure 244. Lone Mountain turquoise. Cuff bracelet. Silver setting. Navajo, signed: N. 1-1/2" x 3". *$400-600.*

Figure 245. Lone Mountain, Chinese and Morenci turquoise. Rings. Gold and silver settings. *$200-4000.*

Figure 246. Lone Mountain, Red Mountain, Fox and Hachita turquoise. Rings. Gold and silver settings. *$100-2200.*

Figure 247. Lone Mountain turquoise. Ring. Gold setting. 1-1/2" x 3/4". *$2000-2400.*

Manassa (Colorado) mined turquoise can be found in South Central Colorado. One of the few notable Colorado mines, Manassa, also known as King's Mine, named for the family currently working the rock, has revealed turquoise of blue-green to green with a lovely yellow to brown matrix. This is one of the truly unique turquoise gems to still be in production today. [Figures 248-249]

Figure 248. Manassa turquoise. Cuff bracelet. Silver setting. Navajo. 1-3/4" x 3". *$450-650.*

Figure 249. Manassa turquoise. Bolo tie. Silver setting with silver tips. Navajo. 4-1/2" x 3". *$400-600.*

Mojave (Montana) is a beautiful and difficult to acquire green turquoise, often referred to as Green Apple turquoise. It is a good quality material and one that takes a nice polish. It is exceptional when set in silver and is truly one of the more unusual turquoise materials for its intense and consistent coloring. [Figures 250-251]

Figure 250. Mojave turquoise. Belt buckle, multiple stones. Silver setting. Signed: John Renner. 5" x 4". *Courtesy of the artist. $1000-1300.*

Figure 251. Mojave turquoise. Belt buckle, multiple stones. Silver setting. Signed: John Renner. 5" x 4-1/2". *Courtesy of the artist. $450-650.*

Morenci (Arizona) is one of the most well-known turquoise names in the world. Ranking in recognition with Kingman and Bisbee, also Arizona mines. Morenci turquoise is quite distinguishable for its pyrite inclusions that, when polished, often take on a silver sheen. Most Morenci turquoise runs from a beautiful dark sky blue to light blue. Many know of Morenci turquoise because it retains the distinction of being the first American mined turquoise to come to market. Worked by many craftsmen into Native American jewelry, Morenci can be found in some stunning older pieces, and though the mine is now depleted, you can still find contemporary artisans working the material, making it highly collectible. [Figures 252-263]

Figure 252. Morenci turquoise. Assorted natural blue-green stones. Eighty-four grams. *Courtesy of John Renner.*

Figure 253. Morenci turquoise. Assorted natural #1 green stones. Sixty grams. *Courtesy of John Renner.*

Figure 254. Morenci turquoise. Pendant, multiple stones. Silver setting. Navajo. 3" x 2-1/4". *$175-300.*

Figure 255. Morenci turquoise. Belt buckle. Silver setting. Navajo, signed: RNS. 3" x 2". *$175-300.*

Figure 256. Morenci turquoise and coral. Necklace, multiple stones, single strand coral beads. Pueblo. 28" length. *$300-450.*

Figure 257. Morenci turquoise. Squash blossom necklace, multiple stones. Silver setting. Navajo, signed: Rajh. 26" length. *$1200-1600.*

Figure 259. Morenci turquoise. Watch band, multiple stones. Silver setting with stretch band. Navajo. 5" length. *$175-300.*

Figure 258. Morenci turquoise. Earrings. Silver setting. Navajo. 1" x 1-1/8". *$150-300.*

Figure 261. Morenci turquoise and coral. Cuff bracelet, multiple stones. Silver setting. Navajo, signed: J.W.C. 2" x 3-1/2". *$450-650.*

Figure 260. Morenci turquoise. Bracelet, multiple stones. Silver setting. Navajo, signed: MA. 1-1/2" x 3". *$200-350.*

Figure 262. Morenci turquoise. Cuff bracelet. Silver setting. Navajo, signed: Tom. 2-1/2" x 3". *$600-800.*

Figure 263. Morenci and Sleeping Beauty turquoise. Rings. Silver settings. *$250-600.*

No. 8 (Nevada) was first mined around 1930, though it has long been depleted. During its height, this mine produced some of the largest turquoise nuggets ever revealed. The No. 8 material is most often found as a light blue with spider webbing of yellow-brown to black, making for beautifully contrasting gem grade stones. [Figures 264-273]

Figure 264. #8 turquoise. Assorted cut and polished stone. Twenty grams. *Courtesy of John Renner.*

Figure 265. #8 turquoise. Belt buckle, multiple stones. Silver setting. Navajo. 2-1/2" x 2-3/4", buckle. *$1600-2000.*

Figure 266. #8 turquoise. Cuff bracelet. Gold and silver setting. Apache, signed: Marc Antia. 1-1/2" x 3". *$1000-1500.*

Figure 267. #8 turquoise. Cuff bracelet. Silver setting. Navajo. 2" x 3". *$450-650.*

Figure 268. #8 turquoise. Cuff bracelet, multiple stone cluster. Silver setting. Navajo. 3-1/2" x 3". *$500-700.*

Figure 269. #8 turquoise. Cuff bracelet, multiple stones. Silver setting. Signed: John Renner. 1-1/2" x 3". *Courtesy of the artist. $500-700.*

Figure 270. #8 turquoise. Cuff bracelet. Silver setting. Signed: John Renner. 3" x 3". *Courtesy of the artist. $450-600.*

Figure 271. #8 turquoise. Pendant. Silver setting. Navajo. 2" x 1-3/4". *$150-300.*

Figure 272. #8 turquoise. Pendant. Silver setting. Navajo, signed: E. 2' x 1-1/2". *$100-200.*

Figure 273. #8 turquoise. Pendant. Silver setting. Navajo, signed: E. 3-1/4" x 1-1/4". *$250-350.*

Orvil Jack (Nevada) is a Nevada mine named for its discoverer, Orvil Jack. Located in the Blue Ridge of the Crescent Valley, Orvil Jack continues to be operated by Jack's daughter. While little material is being extracted from the mine, the turquoise is highly sought after by collectors. Orvil Jack turquoise has the rare yellow-green coloring that comes from its zinc content. If you find yourself in Northern Nevada consider stopping by to see the mine itself. Be sure to check first as, like most mines, even those depleted, they most often rest on private property. [Figures 274-275]

Figure 274. Orvil Jack turquoise. Cuff bracelet, multiple stones. Silver setting. Navajo. 1-1/2" x 3". *$550-700.*

Figure 275. Orvil Jack and Chinese turquoise. Cuff bracelet, multiple stones. Silver settings. Chinese turquoise, left; Orvil Jack turquoise, right. Signed: Tom Dewit, left; Navajo, right. 1" x 3", 1-1/4" x 3". *$500-1200.*

Paiute (Nevada) turquoise is located in Central Nevada, sharing a mountain with the Godber and Burnham mines. Having been in active production only as far back as 1992, the Paiute is one of the regions currently active mines. Though claims at the site date back to the mid-1970s, the limited material coming from Paiute makes it quite desirable. Most of the turquoise is a high-grade material in shades of blue with spider webbing that ranges from black to brown and red to orange. Being a hard stone, Pauite is more readily cut and polished than some other lesser mine grades.

Persian (Iran) turquoise deposits have been mined for over 2000 years. South of the Caspian Sea, and therefore close to trade routes, the turquoise mined from Persia was and remains the finest in the world. Over the course of thousands of years the fine and intense blue turquoise found in Persia, and the term "Persian Turquoise," has been synonymous with the finest quality. The term today is more often meant to be a definition of quality than a statement of origin. Its beautiful variant sky blue color is almost always without matrix and often what many think of as "true" turquoise. While the materials from various mines in the region exhibit a variety of blue colors, Persian turquoise is generally one of the easier localities to identify because of its distinctive color and excellence. [Figures 276-280]

Figure 276. Persian turquoise. Assorted cut and polished stones. Twenty-eight grams. *Courtesy of John Renner.*

Figure 277. Persian turquoise. Watch cuff bracelet. Silver setting. Navajo. 1-1/4" x 3". *$250-350.*

Figure 278. Persian turquoise. Cuff bracelet. Silver setting. 2" x 3". *$450-600.*

Figure 279. Persian turquoise. Earrings, multiple stone cluster. Silver setting. Zuni, signed: Wayne Carla. 3-1/2" x 1-1/4". *$400-500.*

Figure 280. Persian turquoise and coral. Bolo tie. Silver setting with silver tips. Navajo. 3-1/4" x 2-1/4". *$400-600.*

Red Mountain (Nevada) is one of a number of turquoise mines in Lander County, Nevada. Having produced a large quantity of gem grade material, Red Mountain turquoise is the rival of some of the finest turquoise stones of the American Southwest. Often used by some of today's leading Native American artisans, the red spider webbing gives this turquoise its distinctive name and place in any collection. [Figures 281-282]

Figure 281. Red Mountain turquoise. Assorted cut and polished stones. Six grams. *Courtesy of John Renner.*

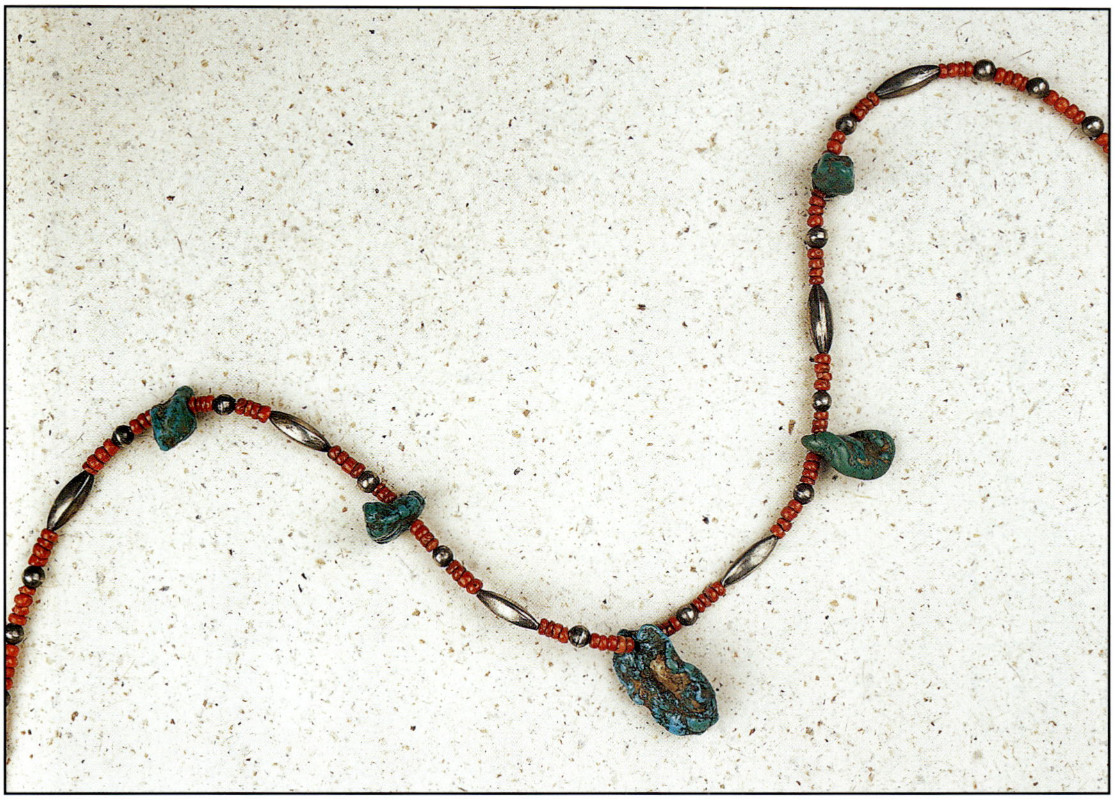

Figure 282. Red Mountain turquoise and coral. Necklace, multiple stone with silver beads. Navajo. 26" length. *$450-650.*

Royston (Nevada) area consists of three mines in Nevada. They are the Bunker Hill, Oscar Wehrend, and the most notable, Royal Blue. Royston district turquoise continues to produce gem grade quality material that range from dark green to a beautiful light blue, often with brown matrix. Because the turquoise is relatively soft, it is highly collectible as the amount of workable material is less than other mines of harder stone. [Figures 283-286]

Figure 283. Royston turquoise. Cut and polished stone. Twelve grams. *Courtesy of John Renner.*

Figure 284. Royston turquoise. Necklace, multiple stone with shell (Wampum) beads. Pueblo. 26" length. *$600-850.*

Figure 286. Royston turquoise. Cuff bracelet, multiple stones. Silver setting. Navajo, signed: C MTZ. 1-1/2" x 3". *$750-950.*

Figure 285. Royston turquoise. Pendant. Silver setting. Navajo, signed: DH. 3-1/2" x 1-3/4". *$250-350.*

Sleeping Beauty (Arizona) is a favorite mined turquoise of the Zuni Pueblo Native Americans. Well-known for its blue color that is most often quite even throughout the stone and devoid of matrix, Sleeping Beauty turquoise is frequently found in petit point and inlay jewelry. Collectors can more readily acquire excellent examples of Sleeping Beauty turquoise because it is one of the largest North American turquoise mine operations still actively being worked today. [Figures 287-346]

Figure 287. Sleeping Beauty turquoise. Assorted rough stones. 186 grams. *Courtesy of John Renner.*

Figure 288. Sleeping Beauty turquoise. Assorted cut and polished stones. Ten grams. *Courtesy of John Renner.*

Figure 289. Sleeping Beauty turquoise, coral, shell, and malachite. Belt buckle, multiple stones. Silver setting. Zuni, signed: Sammy & Esther Anardian. 2-1/2" x 2". *$450-600.*

Figure 290. Sleeping Beauty turquoise. Belt buckle, multiple stone cluster. Silver setting. Signed: John Renner. 3" x 2". *Courtesy of the artist. $350-500.*

Figure 291. Sleeping Beauty turquoise, shell, and black onyx. Belt buckle, multiple stone inlay. Silver setting. Zuni, signed: tj. 2-3/4" x 1-3/4". *$400-500.*

Figure 292. Sleeping Beauty turquoise. Belt buckle, multiple stones. Silver setting. Navajo. 3-1/2" x 3". *$200-300.*

Figure 293. Sleeping Beauty turquoise, shell, coral, and black onyx. Belt buckle, multiple stone inlay. Silver setting. Zuni, signed: Russell Sam. 3" x 2". *$350-500.*

Figure 294. Sleeping Beauty turquoise, agate and mother-of-pearl. Belt buckle, multiple stones. Silver setting. Zuni, signed: W. 3-1/2" x 2-1/2". *$500-650.*

Figure 295. Sleeping Beauty turquoise, shell, and black onyx. Belt buckle, multiple stones, inlay. Silver setting. Navajo, signed: TJ. 3" x 2". *$350-500.*

Figure 296. Sleeping Beauty turquoise, shell, coral. Belt buckle, multiple stones, cluster. Silver setting. Zuni. 3" x 2-1/2". *$275-450.*

Figure 297. Sleeping Beauty turquoise, shell, coral, and black onyx. Belt buckle, multiple stones, inlay. Silver setting. Zuni, signed: Leander and Lisa Othole. 3" x 2". *$400-600.*

Figure 298. Sleeping Beauty turquoise, shell, coral, and black onyx. Belt buckle, multiple stones, inlay. Silver setting. Zuni, signed: RIH. 3-1/2" x 3". *$675-900.*

Figure 299. Sleeping Beauty turquoise. Belt buckle, multiple stone inlay. Silver setting. Navajo. 3-1/2" x 3". *$450-600.*

Figure 300. Sleeping Beauty turquoise. Cross pendant, multiple stone inlay. Silver setting. 4" x 2". *$150-300.*

Figure 301. Sleeping Beauty turquoise. Pendant and pins, multiple stone petit point. Silver setting. Zuni, signed: Penteah. 3", 2-1/2", 1". *$100-300.*

Figure 302. Sleeping Beauty turquoise, coral, shell, and black onyx. Pin, multiple stones, inlay. Silver setting. Zuni, signed: G. Smith and Ray Watson. 1-1/4" x 1". *$200-350.*

Figure 303. Sleeping Beauty turquoise. Pin, multiple stones. Silver setting. Navajo, signed: JR. 3". *$150-300.*

Figure 304. Sleeping Beauty turquoise. Pin, multiple stones, cluster. Silver setting. Zuni. 1-1/4" x 1". *$80-150.*

Figure 305. Sleeping Beauty turquoise. Pin, multiple stone squash blossom. Silver setting. Navajo. 2" x 1". *$80-120.*

Figure 306. Sleeping Beauty turquoise. Pin, multiple stones, cluster center. Silver setting. Navajo. 6" x 4". *$600-900.*

Figure 307. Sleeping Beauty turquoise. Pin, multiple stone cluster. Silver setting. Zuni. 3-1/2". *$800-1100.*

Figure 309. Sleeping Beauty turquoise and coral. Pendant, multiple stones. Silver setting. Zuni, signed: Effie C. 2" x 1". *$75-150.*

Figure 308. Sleeping Beauty turquoise. Pendant, multiple stones. Silver setting. Zuni, signed: Effie C. 2-1/2" x 1-1/2". *$150-250.*

Figure 310. Sleeping Beauty turquoise, coral, shell, and black onyx. Pendant with beads, multiple stone inlay. Silver setting. Zuni, signed: MS. 18" length. *$400-600.*

Figure 311. Sleeping Beauty turquoise. Necklace, multiple stone cluster. Silver setting. Zuni. 20". *$150-250.*

Figure 312. Sleeping Beauty turquoise. Necklace, multiple stones. Silver setting. Zuni. 18" length. *$250-375.*

Figure 313. Sleeping Beauty turquoise. Necklace, multiple stone inlay. Silver setting, beads. Navajo. 18" length. *$900-1400.*

Figure 314. Sleeping Beauty turquoise, shell, and coral. Necklace, multiple stones. Zuni. 24" length. *$250-350.*

Figure 315. Sleeping Beauty turquoise. Belt buckle, multiple stone cluster. Silver setting. Navajo. 3-1/2" x 3". *$200-300.*

Figure 316. Sleeping Beauty turquoise. Belt Buckle, multiple stones. Silver setting. Zuni, signed: Effie C. 3-1/2" x 3". *$375-500.*

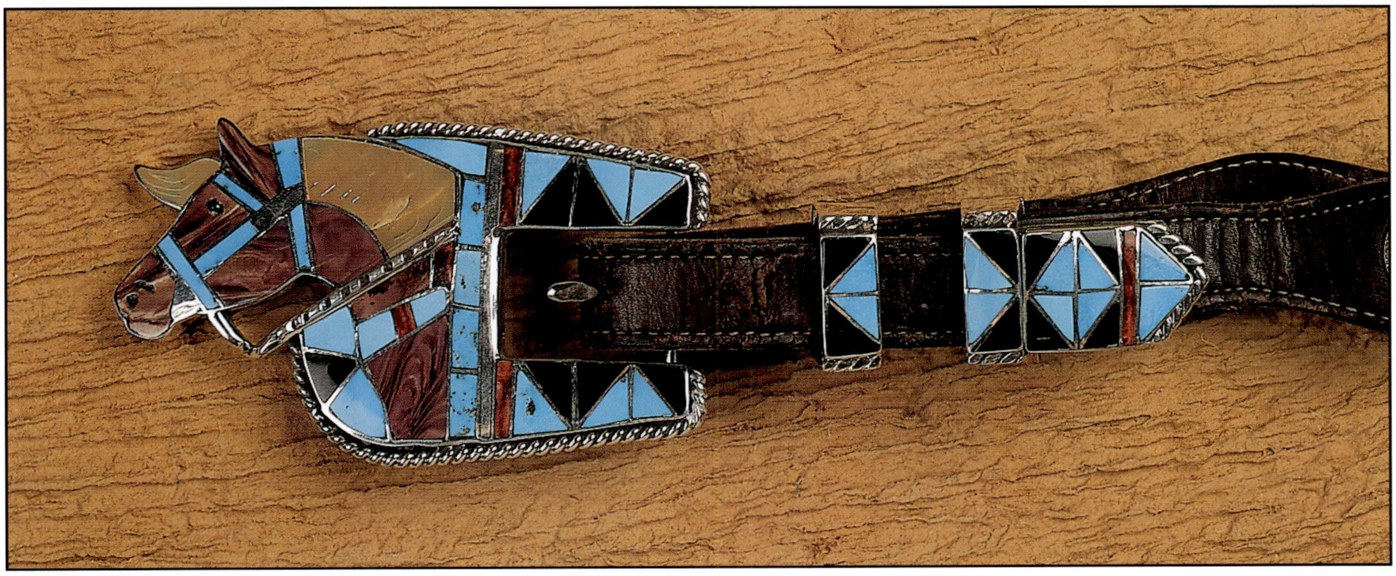

Figure 317. Sleeping Beauty turquoise, coral, agate, and shell. Belt buckle, multiple stone inlay. Silver setting. Zuni, signed: H-L. 4" x 2-1/2". *$500-750.*

Figure 318. Sleeping Beauty turquoise. Earrings, multiple stones. Silver setting. Zuni. 1". *$90-150.*

Figure 319. Sleeping Beauty turquoise. Earrings, multiple stones. Silver setting. Zuni, signed: Effie C. 1-3/4" x 3/4". *$175-275.*

Figure 320. Sleeping Beauty turquoise, coral, shell, and black onyx. Earrings, multiple stone inlay. Silver setting. Zuni. 1" x 1". *$80-120.*

Figure 321. Sleeping Beauty turquoise. Earrings, multiple stones. Silver setting. Zuni, signed: Effie C. 1" x 1/2". *$75-150.*

Figure 322. Sleeping Beauty turquoise, coral, shell, and black onyx. Earrings, multiple stone inlay. Silver setting. Zuni. 1" x 1". *$75-125.*

Figure 323. Sleeping Beauty turquoise. Earrings, multiple stone inlay. Silver setting. Navajo. 1". *$80-120.*

Figure 324. Sleeping Beauty turquoise, coral, shell, and black onyx. Earrings, multiple stone inlay. Silver setting. Navajo. 3/4" x 1/2". *$80-120.*

Figure 325. Sleeping Beauty turquoise, coral, shell, and black onyx. Bolo tie, multiple stone inlay, and silver tips. Silver setting. Zuni, signed: Dev Eriacho. 2-1/4" x 2". *$300-400.*

Figure 326. Sleeping Beauty turquoise, coral, shell, black onyx, and sugilite. Bolo tie, multiple stone inlay, and silver tips. Silver setting. Zuni. 1-3/4" x 1-1/4". *$150-200.*

Figure 327. Sleeping Beauty turquoise, coral, shell, and black onyx. Bolo tie, multiple stone inlay, and silver tips. Silver setting. Zuni. 2" x 2-1/4". *$125-250.*

Figure 328. Sleeping Beauty turquoise, coral, and lapis. Bolo tie, multiple stone inlay, and silver tips. Silver setting. Navajo. 2-1/2" x 2-1/2". *$200-400.*

Figure 329. Sleeping Beauty turquoise, coral, shell, lapis, and black onyx. Bolo tie, multiple stone inlay, and silver tips. Silver setting. Zuni. 6" x 4". *$500-750.*

Figure 330. Sleeping Beauty turquoise, coral, shell, and black onyx. Bolo tie, multiple stone inlay. Silver setting. Zuni, signed: Roger Cellicion. 6" x 4". *$2000-3000.*

Figure 331. Sleeping Beauty turquoise. Watch cuff bracelet, multiple stones. Silver setting. Navajo. 3-1/2". *$200-300.*

Figure 332. Sleeping Beauty turquoise. Cuff bracelet, multiple stone cluster. Silver setting. Navajo, signed: F M Begay. 4" x 3". *$700-900.*

Figure 333. Sleeping Beauty turquoise. Cuff bracelet. Silver setting. Navajo, signed: Roger Nelson. 3-1/2" x 1-3/4". *$350-500.*

Figure 334. Sleeping Beauty turquoise and coral. Cuff bracelet, multiple stones. Silver setting. Zuni, signed: Effie C. 3" x 1". *$375-500.*

Figure 335. Sleeping Beauty turquoise. Cuff bracelet, multiple stones. Silver setting. Navajo, signed: RC. 3" x 3". *$450-650.*

Figure 336. Sleeping Beauty turquoise. Cuff bracelet, multiple stone inlay. Silver setting. Navajo. 3" x 1-1/4". *$450-650.*

Figure 337. Sleeping Beauty turquoise. Cuff bracelet, multiple stones. Silver setting. Navajo. 4" x 3". *$800-1100.*

Figure 338. Sleeping Beauty and Kingman turquoise with coral. Choncha belt, multiple stone cluster. Silver setting. Navajo. 2-3/4". *$1800-2400.*

Figure 339. Sleeping Beauty turquoise. Choncha belt, multiple stone cluster buckle. Silver setting. Navajo. 3-1/2" x 3-1/2", 3" x 2-3/4", 2" x 1-3/4". *$6000-7500.*

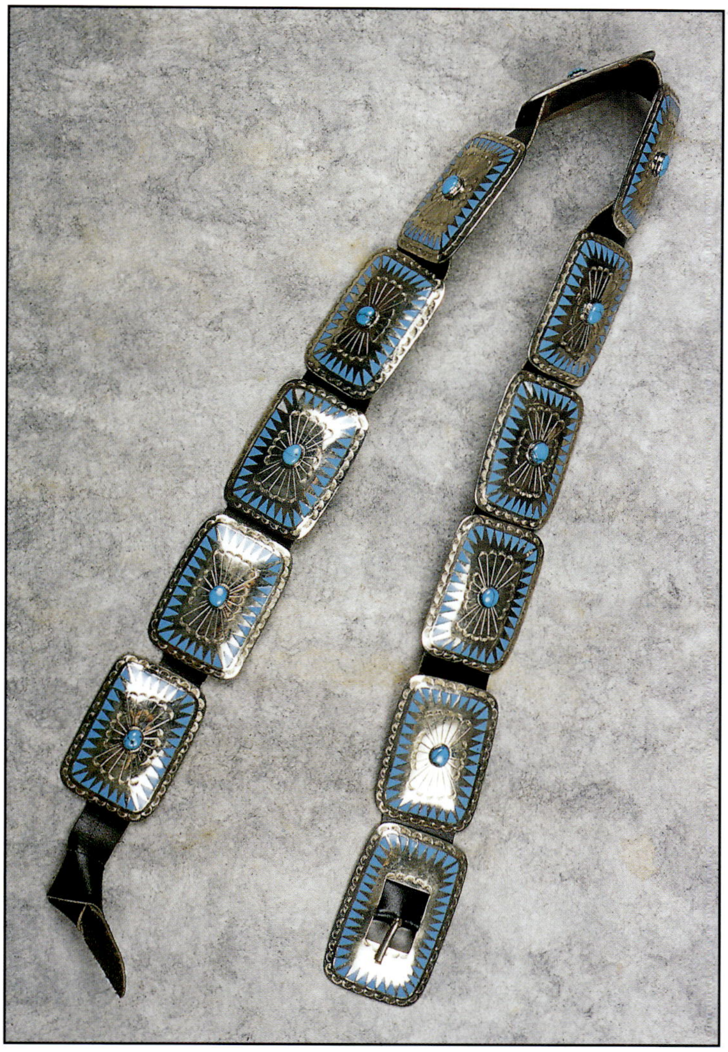

Figure 340. Sleeping Beauty turquoise. Choncha belt, multiple stones. Silver setting. Navajo. 3-1/2" x 2-3/4". *$1850-2250.*

Figure 341. Sleeping Beauty turquoise, coral, shell, and black onyx. Rings, multiple stones; left, chip inlay. Silver setting. Navajo. 1-1/4" x 1-1/4", 2" x 1". *$100-400.*

Figure 342. Sleeping Beauty turquoise, coral, shell, and black onyx. Rings, multiple stones, inlay. Gold setting, left; silver setting, right. Zuni. 1" - 1-1/4". *$250-700.*

Figure 343. Sleeping Beauty turquoise. Cuff bracelet, multiple stone cluster. Silver setting. Navajo, signed: Verdy J. 3" x 1-1/2". *$600-850.*

Figure 344. Sleeping Beauty turquoise. Watch band, multiple stone inlay. Silver setting. Navajo. 6" x 1". *$250-350.*

Figure 345. Sleeping Beauty turquoise. Sculpture. Silver. Navajo, signed: Wilford Begay. 9" height. *$2000-3000.*

Figure 346. Sleeping Beauty turquoise. Sculpture. Silver. Navajo, signed: Wilford Begay. 9" height. *$2000-3000.*

Stennich (Nevada) mine turquoise is one of the few mines that predominates turquoise of green shades. A wide array of greens can be found in Stennich turquoise that make beautifully set pieces when cut and polished. It is a hard, high-grade material ideal for use alone or in association with other turquoises and gems. [Figures 347-348]

Figure 347. Stennich turquoise. Belt buckle, multiple stone cluster. Silver setting. Navajo, signed: CY. 1-3/4" x 1-1/2". *$200-350.*

Figure 348. Stennich turquoise. Belt buckle, multiple stone cluster. Silver setting. Navajo, signed: CY. 2" x 1-3/4". *$200-350.*

Stormy Mountain (Nevada) turquoise is distinctive for the dark blue stones mined here. This material will often include a blotchy, black chert matrix resembling storm clouds, whereby the mine got its name. Stormy Mountain turquoise can be confused with the similar looking material mined from Blue Diamond. [Figures 349-350]

Figure 349. Stormy Mountain turquoise, coral, and black onyx. Necklace, multiple stones inlay with beads. Silver setting. Navajo. 26" length. *$300-475.*

Figure 350. Stormy Mountain turquoise. Pendant. Silver setting. Navajo. 6" x 1-1/2". *$600-800.*

Timberline (Nevada) mine is one of Nevada's smallest worked turquoise mines. The high-grade material from Timberline was small in quantity and ranges from medium to dark blue in color, often with brown to black matrix. A rather difficult material to attain, it is a natural addition to a comprehensive collection. [Figures 351-352]

Figure 351. Timberline turquoise. Assorted cut and polished stone. Fourteen grams. *Courtesy of John Renner.*

Figure 352. Timberline turquoise. Cuff bracelet, multiple stone. Silver setting. Signed: John Renner. 1-1/4" x 3". *Courtesy of the artist. $650-850.*

Turquoise Mountain (Arizona) near the famed Kingman mine was closed relatively recently, in the late 1980s. The material from this mine is a lovely light to dark blue and can be found with or without spider webbing. Often referred to as "Birdseye," the mined turquoise is one and the same. Birdseye refers to turquoise from the mine that shows an area of blue circled by a darker blue matrix, thereby resembling a bird's eye. A cut and polished piece from Turquoise Mountain is an outstanding addition to any collection. [Figures 353-357]

Figure 353. Turquoise Mountain turquoise. Assorted rough stones. 238 grams. *Courtesy of John Renner.*

Figure 354. Turquoise Mountain turquoise. Cut and polished stone. Four grams. *Courtesy of John Renner.*

Figure 355. Turquoise Mountain turquoise. Belt buckle, multiple stones. Silver setting. Navajo. 5" x 4". *$850-1100.*

Figure 356. Turquoise Mountain turquoise. Cuff bracelet. Silver setting. Navajo. 3-1/2" x 3". *$550-750.*

Figure 357. Turquoise Mountain turquoise. Pendant, multiple stones. Silver setting. Navajo. 2-1/2" x 1". *$250-375.*

Tyrone (New Mexico) turquoise has suffered a sad fate. While the mine was associated with the more valued copper mining operations southwest of Silver City, material mined from Tyrone is an intense high-grade blue stone. Unfortunately, no turquoise has been mined from Tyrone since the early 1980s. It was at this time that the method of extracting the copper ore was processed through an acid wash rather than the more widely used crushing operation. The use of acid destroys any of the turquoise to be found in the copper ore. A well-known mineral streak begins east of Silver City, New Mexico, and wends its way through Arizona and into Mexico. This streak includes many outcrops and mines of various well-known Southwest minerals, including the turquoise of Morenci. [Figures 358-360]

Figure 358. Tyrone turquoise. Belt buckle. Silver setting. Navajo, signed: NB. 3-1/4" x 3". *$350-450.*

Figure 359. Tyrone turquoise. Watch bracelet, multiple stones, inlay. Silver setting. Navajo. 1" x 3". *$250-350.*

Figure 360. Tyrone turquoise. Cuff bracelet. Silver setting. Navajo. 3" x 3". *$1500-2000.*

South American turquoise has been mined for centuries from a number of areas in the western portion of the continent. The material is not easily distinguishable from North American turquoise, and is most often identified through provenance. It is an important historical source for the collector because of the early mining that was done. Though little information is available on the individual mines or countries where material has come to market. [Figures 361-362]

Figure 361. Peruvian turquoise. Choncha belt, multiple stones. Silver setting. Signed: John Renner. 4" x 3-1/2", buckle; 3-3/4" x 3", Choncha. *Courtesy of the artist. $4000-6000.*

Figure 362. Peruvian turquoise. Belt buckle. Silver setting. Signed: John Renner. 6" x 4". *Courtesy of the artist. $700-900.*

Variscite (Utah) is a rare cousin of turquoise and is found in some instances in Utah, in the American Southwest. While not turquoise by chemical composition, variscite is a very hard stone and on occasion has a blue to green coloring giving the look of natural turquoise. While Utah mines of variscite are pretty well depleted today, for a number of years a fair amount of material was released into the marketplace. Because of its proximity to the color, degree of matrix, and webbing, variscite became a favorite of jewelers and silversmiths, thus discussed and illustrated here. While it is difficult to find a good piece of variscite used in Southwestern jewelry today, it is an important adjunct mineral to add to a comprehensive collection, and looks beautiful cut, polished, and set in its own right. [Figure 363]

Figure 363. Variscite. Pendant. Silver setting. Navajo, signed: E. 2-1/2" x 1-3/4". *$90-150*

Figure 364. Reconstituted turquoise. Belt buckle, multiple stone cluster. Silver setting. Navajo. 4" x 3". *$200-300.*

Figure 365. Reconstituted turquoise. Necklace, five strand bead. Navajo. 24" length. *$400-600.*

Figure 366. Reconstituted turquoise. Pendants, multiple stone cluster. Silver setting. Navajo. 1", 2" x 1-1/2", 3". *$100-300.*

Figure 367. Reconstituted Kingman turquoise. Fetish necklace, multiple stones. Navajo. 18" length. *$150-250.*

Figure 368. Reconstituted turquoise. Bolo tie, multiple stone cluster. Silver setting. Navajo. 3-1/4". *$250-350.*

Figure 369. Reconstituted turquoise. Bolo tie, multiple stone cluster. Silver setting and tips. Navajo. 2-1/2" x 2". *$75-150.*

Figure 370. Reconstituted turquoise, coral, shell, and black onyx. Cuff bracelet, multiple stones. Silver setting. Navajo. 2-1/4" x 3". *$225-300.*

Figure 372. Block turquoise. Earrings, multiple stones. Navajo. 3-1/4" x 1-1/2". *$15-25.*

Figure 371. Reconstituted turquoise. Cuff bracelet, multiple stone inlay. Silver setting. Navajo. 1-1/2" x 3". *$200-300.*

Figure 373. Block turquoise. Earrings, multiple stone inlay. Silver setting. Navajo. 3/4" x 5/8". *$25-50.*

Photography credits have been listed in each caption. Schiffer Publishing, Ltd. has provided any unlisted photographic figure illustrations with the approval of the author. All pieces not otherwise credited are provided courtesy of JR's Southwestern Trading Post.

Glossary

Agate: Fine-grained variety of chalcedony, often with colored bands or irregular splotches.

Aluminum: Silvery-white, metallic element, the most abundant in the earth's crust, though found only in combination, chiefly in bauxite.

Amazonite: Green variety of microcline, often used as a semiprecious stone.

American Southwest: A region of the United States, often drier in weather than other United States regions.

Anazasi: Otherwise known as Pueblo. Name given by the Spanish to Native Americans who lived in adobe or stone houses in the Southwest United States.

Apatite: Common complex mineral consisting of calcium fluoride phosphate or calcium chloride phosphate; a source of phosphorus.

Aztec: People of central Mexico whose civilization reached its height at the time of the Spanish conquest in the early sixteenth century.

Bolo tie: Necktie of thin cord fastened in front with a decorative clasp.

Cabachon: Any precious stone cut and polished, but not faceted, of convex hemispherical or oval form.

Chalcedony: Microcrystalline, translucent variety of quartz, often appearing as milky or grayish in color.

Chert: Heavily compacted rock consisting nearly entirely of microcrystalline quartz.

Chrysocolla: Hydrous copper silicate mineral, occurs in compact green or blue masses.

Coin silver: Metal containing 90 parts silver and 10 parts other metal. Made of melted down pre-1900 Mexican and United States coins.

Color Enhanced: Chemically altered to change the color of a stone.

Color Treated: Often used term meaning color enhanced.

Concha: Hammered silver that often resembles a shell, flower, or sunburst; often having designs.

Coral: Hard, variously colored, calcareous skeleton secreted by certain marine polyps.

Copper: Metallic element having a distinctive reddish-brown color.

Doublet: Composite, or assembled from two or more components.

Gemstone: Any precious or semiprecious stone that can be cut and polished for use as a gem.

Gold: Bright and precious yellow metallic element not subject to oxidation or corrosion.

Hohokam: An American Indian culture of the central and southern deserts of Arizona, approximately 450–1450 C.E., roughly at the same time with the Anasazi culture to the north.

Imitation: Result or product of imitating; copy.

Inlay: Decorate an object with layers of materials set into the surface.

Intarsia: An art or technique of decorating a surface with inlaid patterns.

Iron oxide: Any one of various oxides of iron, such as ferric oxide or ferrous oxide.

Jade: Either of two minerals, jadeite or nephrite, usually green. Often used in jewelry or carved.

Jasper: Compact, opaque, cryptocrystalline variety of quartz, most often colored red.

Lapidary: The process of cutting and polishing gems and minerals.

Lapis Lazuli: Deep-blue mineral composed mainly of lazurite with smaller quantities of other minerals.

Matrix: The portion of a rock in which coarser crystals or rock fragments are surrounded.

Mimetite: A mineral that occurs in pale yellow or brownish hexagonal crystals.

Mine: Place in or on the Earth where minerals may be extracted.

Mohs Scale: Scale of hardness used in mineralogy, from softest, 1 (talc), to hardest, 10 (diamond).

Mother-of-pearl: Hard, iridescent substance that forms the inner layer of certain mollusk shells.

Navajo: Largest tribal group in the United States, located in Arizona and New Mexico.

Persia: Ancient empire located in Asia extending from Egypt and the Aegean to India; conquered by Alexander the Great, 334-331 B.C.E.

Petit point: A stone shaped to a fine point that is often larger than needle point, characterized by being round, oval, or having one rounded end.

Pyrite: Common brass-yellow iron disulfide mineral having a metallic luster. Often referred to as "Fool's Gold".

Pyromorphite: Mineral; lead chlorophosphate that occurs in crystalline and massive forms, most often a green, yellow, or brown color.

Onyx: Variety of chalcedony having straight parallel bands of alternating colors; often referred to as "Jet" when black.

Reconstituted: Pulverizing pieces of stone that are then stabilized and hardened with resins to achieve a natural appearance.

Silver: White metallic element.

Sphalerite: Common mineral, zinc sulfide, typically containing iron and cadmium, occurring in yellow, brown, or black crystals or cleavable masses.

Spider web: Name for pattern, often even, of matrix in turquoise stone.

Stabilized: Infusing natural stone with a clear epoxy resin, under pressure permanently hardens the rock and deepens the color.

Squash blossom: Design or configuration akin to the flower of the squash plant.

Sutter's Mill: The location in California, near Sacramento, where gold was discovered in 1848, precipitating the gold rush of 1849.

Tiffany: Charles Lewis, 1812-1902, U.S. jeweler. Louis Comfort, 1848-1933, U.S. painter and decorator, especially of glass. Also known for high quality retail goods with distinctive blue color packaging.

Turquoise: Hydrous phosphate of aluminum and copper used as a gem. It occurs rarely in crystal form, but is usually cryptocrystalline. Turquoise is opaque and often has a waxy luster; the color varies from pale blue to sky-blue, dark blue, green, white, light brown.

Variscite: Secondary hydrated phosphate of aluminum mineral, occurring mainly as substantial bluish-green nodules.

Vein: Area in host rock in which gem material is found; formed by seeping water in open area of rock.

Wampum beads: Cylindrical beads made from shells, pierced and strung.

Zinc: Bluish-white metallic element.

Zuni: Native North American Indians inhabiting the largest Indian pueblos, in western New Mexico.

Bibliography

Chalker, Kari. *Totems to Turquoise: Native North American Jewelry Arts of the Northwest and Southwest*. New York, New York: Harry N. Abrams, 2004.

Dubin, Lois Sherr. *North American Indian Jewelry and Adornment*. New York, New York: Harry N. Abrams, 1999.

Foxx, Jeffrey Jay and Carol Karasik. *Turquoise Trail: Native American Jewelry and Culture of the Southwest*. New York, New York: Harry N. Abrams, 1993.

Holden, Martin. *The Encyclopedia of Gemstones and Minerals*. New York, New York: Michael Friedman Publishing Group, Inc., 1991.

Johnsen, Ole. *Princeton Field Guide. Minerals of the World*. Princeton, New Jersey: Princeton University Press, 2002.

Kolbe, Regina. "Tribal Jewelry Hot Again with Western Look." *Antique Trader*, August 10, 2005.

Kuhns, Elizabeth. "Wild About Turquoise." *Lapidary Journal*, 2003.

Lowry, Joe Dan. *Turquoise Unearthed: An Illustrated Guide*. Santa Fe, New Mexico: Rio Nuevo Publishers, 2002.

Pough, Frederick H. *A Field Guide to Rocks and Minerals, 5th edition*. New York, New York: Houghton Mifflin, 1998.

Schiffer, Nancy. *Turquoise Jewelry, 2nd edition*. Atglen, Pennsylvania: Schiffer Publishing Ltd., 2004.

____. *Jewelry by Southwest American Indians: Evolving Designs*. Atglen, Pennsylvania: Schiffer Publishing Ltd., 1997.

Simmons, Marc. *Turquoise and Six-Guns: The Story of Cerrillos, New Mexico*. Santa Fe, New Mexico: Sunstone Press; 3rd edition, 1989.

Stacey, Joseph. *Arizona Highways*. Phoenix, Arizona: Arizona Department of Transportation, January 1974.

Turnbaugh, William A. and Sarah Peabody Turnbaugh. *Indian Jewelry of the American Southwest*. Atglen, Pennsylvania: Schiffer Publishing, 1997.

Vigil, Arnold (ed.). *The Allure of Turquoise*. Santa Fe, New Mexico: New Mexico Magazine; 2nd edition, 2005.

Zucker, Benjamin. *Gems and Jewels: A Connoisseur's Guide*. New York, New York: Thames and Hudson Publishing Co., 1984.

http://www.bluearticles.com/Article/The-Lure-Of-Turquoise/2115.

http://www.thesantafesite.com/Articles/truthonturquoise .html.

http://encyclopedia.jrank.org/Tum_Van/Turquoise.html.

Index